GROWING RURAL-URBAN DISPARITY IN GUJARAT

GROWING RURAL-URBAN DISPARITY IN GUJARAT

Series Author

S. S. P. Sharma

K. V. Raju

&

Sabyasachi Sarkar

Assoc. Prof., IRMA, Anand

NATIONAL INSTITUTE OF RURAL DEVELOPMENT (NIRD)

RAJENDRANAGAR, HYDERABAD

INSTITUTE OF RURAL MANAGEMENT

SERIALS PUBLICATIONS

NEW DELHI (INDIA)

First Published - 2009

ISBN: 978-81-8387-222-5

Published by

SERIALS PUBLICATIONS
4830/24, Prahlad Street, Ansari Road
Darya Ganj, New Delhi-110002 (India)
Phone : 23245225
Fax : 91-11-23272135
E-mail: serials@satyam.net.in, w.w.w.serialspublications.com

CONTENTS

PREFACE

Negative impacts of liberalization policies were recognized soon after launching of these policies and the need for ameliorating measures was recognized to give it a human face. The processes of globalization and economic liberalization ushered in the early nineties have contributed to the acceleration of economic growth. But they also appeared to have resulted in widening the gap between the rich and the poor. They also seem to be contributing to further regional inequality, family level food insecurity, and unemployment. Liberalization has resulted in a greater degree of skewness in the pattern of development, perhaps leading to the marginalization of some States in the overall dispensation. The problems and constraints associated with agriculture sector seem to have worsened under the reforms regime.

Quality and security of rural livelihood seems to have been adversely affected with the low level of spending by both center and states on social sectors including health, family welfare and education. If access to food, education, technology, credit, insurance, information and natural resources like land, water and forests is going to be mediated by market forces alone, the lot of the economically marginalized with very little purchasing power at their disposal will become more precarious. Both first and second generation reforms so far overlook various social objectives such as eradication of poverty, prevention of concentration of economic power in fewer hands, regional imbalances and promotion of labour intensive economic activities creating more job opportunities.

In the light of the above, a collaborative study on growing rural-urban disparity in Gujarat state was taken up by IRMA which forms a part of a nation wide larger study spearheaded by NIRD (in Bihar and Assam), IGIDR (in Maharashtra), ISEC (in Karnataka), and IEG (in Uttaranchal).

The objectives of the study were:

1. To assess the rural and urban disparity in Gujarat vis-à-vis the disparity in India·
2. To assess the rural and urban disparity across regions and sectors in Gujarat
3. To assess the rural and urban disparity across three different urban areas and the six selected villages in the vicinity of each of them and
4. To suggest the ways, if any, to bridge rural urban disparity.

Section-1 of this report outlines some macroeconomic concerns at all India level followed by assessment of the rural urban disparity based on available secondary data in Gujarat vis-à-vis India in Section-2 and in different regions of Gujarat in Section-3.

Section-4 of this report details the selected urban rural clusters in terms of demographic features, literacy, and amenities available and socioeconomic profile of worker population. Section-5 focuses on the results of the field investigations carried out in the three different urban clusters and the six different villages selected in the vicinity of each of them. Patterns and difference across rural and urban and across clusters were examined. Demographic features like size of the household, sex ratio, children population and literacy status; economic features income across different categories of the sample households and composition of incomes from different sources; and the patterns of employment across rural and urban areas were examined. Availability, access and use of social infrastructure like education, health and drinking water facilities and physical infrastructure like approach roads, transport, communication, electricity and access to market were examined to infer about the dimensions of quality of life. Differences in the quality of services between rural urban areas and across clusters were commented upon the basis of observations made during the field survey.

METHODOLOGY AND STUDY PARAMETEAS

Population figures and sex-wise composition, density of population, sex ratio, and vital rates were chosen to understand the demographic differences across rural and urban areas. Net Domestic Product (NDP), per capita consumption expenditures, distribution of workers and unemployment across rural and urban areas were considered for understanding economic dimensions. Literacy, educational levels and educational status of rural populations, vaccination and health facilities in rural and urban areas were considered to study social dimensions. Rural housing, communication facilities, coverage of public distribution system, type of fuel used for cooking and lighting, access to drinking water and sanitation facilities were considered to indicate quality of life. Section-2 of this report will focus attention on demographic, economic and quality of life dimensions related to Gujarat vis-à-vis India to study the rural-urban disparity.

Per capita Net State Domestic Product (NSDP), sectoral growth rates in income, income, employment and labour productivity were examined to understand macroeconomic trends of development in Gujarat. Work force in Gujarat, availability of skills, type of employment, under employment and unemployment, employment across sectors and women in work force were also examined. Section-3 of this report will concentrate on poverty in the rural and urban areas in the different regions in Gujarat and life expectancy at birth and infant mortality and state expenditures on health in the Gujarat.

The study team decided the selection criteria on the basis of information on population, percentage of people engaged in primary, industrial and services sector for assessing rural urban disparity in three different urban areas and the six selected villages around each. The Worker Populations of all the urban areas in Gujarat were examined. Keeping in view the criteria of the districts where urbanization index was less than 25 according to 1991 Census, the three districts and the urban areas were selected for conducting the study. These three districts also belong to three different regions in Gujarat–South Gujarat, North Gujarat and Saurashtra regions–apart from different agro-climatic zones. The three different urban areas selected for the study also had different sectoral composition in terms of relative importance of primary, secondary and tertiary sectors of the economy.

1. **Predominance of services:** Rajpipla, Narmada District–South Gujarat region; Southern Gujarat agro-climatic region.

As of 1991, Rajpipla town has 3763 enterprises and 16% of total labourers and 23 % of labourers are engaged in agriculture and trade and commerce related activities respectively. There is a possibility that the settlements in the vicinity are dependent on Rajpipla market for selling their produce and buying inputs and consumption goods.

Being a tribal dominated district, there is every possibility that due to lack of infrastructure, the market is concentrated in this town. For conducting the study, the following six villages in the vicinity of Rajpipla Town have been selected:

i. Rampara (Mangrol) ii. Thari
iii. Vavdi iv. Karantha
v. Lachhras vi. Mangrol

2. **Predominance of industrial manufacturing activity:** Thangadh, Surendranagar District-Saurashtra region; North Saurashtra agro-climatic zone.

Field staff of Aga Khan Rural Support Programme (India), Chotila, Surendranagar district assisted collection of primary data in this cluster. Thangadh is famous as center of Ceramic industry in India and basically a manufacturing kind of set up. This town has 1689 enterprises and the maximum number of labourers (58% of total labourers) engaged in manufacturing, processing, servicing and repairs related activities. Also looking at the commodities imported, manufactured and exported, the town seems to be manufacture dominated. For conducting the study, the following six villages in the vicinity of Thangadh Town have been selected:

i. Tarnetar ii. Vijalia
iii. Manadasar iv. Rupavati
v. Songadh vi. Morthala

3. **Predominance of agricultural commodity trading:** Unjha, Mehsana District–North Gujarat region; North Gujarat agro-climatic region.

Faculty and students of North Gujarat Hemchandra University assisted in collection of primary data in this cluster. Unjha town is known for its Ayurvedic Pharmacy engaged in the manufacture of Ayurvedic medicines. The market yard of Unjha is the biggest in the district where a large transaction of jeera (cumin), variyali (fennel seed) and isabgul (fleawort seed) is carried out. Ganj Bazaar at Unjha is the Asia's biggest spice market. Unjha is basically a trading hub kind of set up. This town has 3763 enterprises and the maximum number of labourers (35% of total labourers) engaged in trading and commerce related activities. Also looking at the commodities imported, manufactured and exported, the town seems to be trade dominated. For conducting the study, the following six villages in the vicinity of Unjha Town have been selected:

i. Jagannathpura
ii. Kahoda
iii. Bhunav
iv. Sunak
v. Shihi
vi. Surpura

FOREWORD

Spatial differences or inequalities in poverty between urban and rural areas are commonly found at the global level. However, determining the disparities between rural and urban areas is more difficult than it might seem, since it is difficult to define what is urban and what is rural. Economists define an area as urban when the economy is characterized by non-extractive occupations, for example, industry, commerce, that benefit particularly from a high population density and the accompanying infrastructure. However, improved transport and the relative reduction in transport costs have made it easier to commute between an urban area and its surrounding rural areas or to migrate temporarily from rural to urban areas.

India has a total population of over 1 billion, of which 28% live in urban areas. There has been growth in the percentage of the urban population residing in large cities: it has increased from 60% in early 1980s to over 65% in the early 1990s. Overall, the growth rate of the urban population has slowed from the record level of 3.8% per annum in the 1970s to 3.1% in the 1980s and 2.7% in the 1990s. The slowing down has been greater in smaller towns. The inequality is increasing both in terms of per capita incomes and other socio-economic indicators. In recent decades, however, both the rates of out-migration from backward states and migration into developed states have been reduced. The decline in out-migration from rural areas in backward states like Bihar, Madhya Pradesh, Rajasthan and Uttar Pradesh which have high population growth rates presents a serious problem for their absorption in the rural economy. Seasonal and circular migration of labour for employment has become one of the most durable components of the livelihood strategies of people living in rural areas.

During the post 90s, the rural-urban gap is widening. NIRD has taken a longitudinal study in six states in the first phase, i.e. Assam, Bihar, Gujarat, Karnataka, Maharashtra and Uttrakhand. The second phase study in Madhya Pradesh, Orissa, Tamil Nadu and U.P. is going on. From the first phase study, it is reported that the disparity in social consumption in Karnataka and Uttrakhand is less pronounced than in Bihar and Assam. Surprisingly, the disparity in leading states-Gujarat and Maharashtra is increasing. However, the income-increase in rural area is reported almost in all the states but very slow in Bihar. The study was coordinated by S. S. P. Sharma, Professor and Head, Centre for Water and Land Management (CWLM) and completed by the lead institutes of the country–IGIDR, Mumbai (Maharashtra), IRMA, Anand (Gujarat), ISEC, Bangalore (Karnataka) and IGE, Delhi (Uttrakhand). I hope this volume will provide help to the policy makers at the Centre as well as in the states to explore some mechanism to bridge the emerging gap between rural and urban areas in the country.

V. S. Sampath

V. S. Sampath
Director General

Hyderabad,
January, 2008

1

RURAL URBAN DISPARITY AT ALL INDIA LEVEL

CHANGES IN EMPLOYMENT ACROSS SECTORS

Sectoral contributions to GDP changed significantly over a period of time leaving the employment in different sectors, especially, in agriculture not changed. In 1960 agriculture area employed in parts or in full 74% of people of the country and it reduced to 62% in 1992 and is expected to further fall to 50% of people in agriculture by 2010, whereas, the demand of agricultural products will double in quantity. Productivity using technology and post harvest management will have to compensate the manpower reduction in farming and agricultural products sector.

In the case of industry, in 1960, 11% of the population was employed in small scale and large-scale industries. The trend continued with 11% even during 1992. However, it has to increase by 25% in 2010, as the GDP growth with high technology in the situation of opening up of the economy under WTO region. The pattern of employment will take a new shape. Service with knowledge industry component from 15% employability in 1960 has increased to 27% in 1992. And further it will increase to 50% in view of infrastructure maintenance areas, financial sector, IT

Figure 1

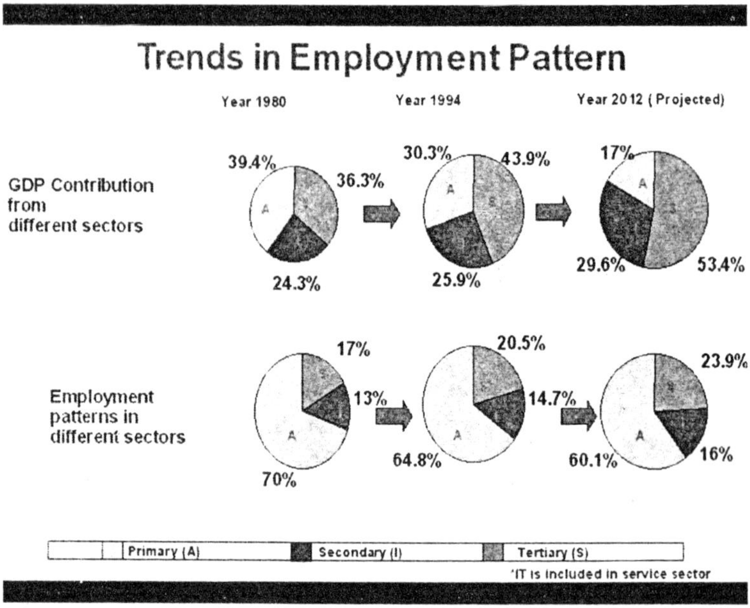

sector and entertainment demands. This big change will demand in all areas more trained skilled human power and technology personnel. Our industrialists, commercial chiefs and technologists may have to get ready for such transformation in agricultural, industries and service-knowledge industries for which human manpower with knowledge and skills have to be evolved in a mission mode (APJ Kalam, in his address at IRMA on 12.04.2002).

IMPACT OF SLOWER AGRICULTURAL GROWTH RATE

What causes great concern now-a-days is the sharp increase in the level of disparities after a few decades of planning, especially because planning was conceived as an instrument to narrow down rural-urban disparities. Rural India encompasses a little less than three-fourths of the country's population and is characterised by low income levels, poor quality of life and a weak base of human development. Nearly one-third of the national income comes from villages, but there is a significant rural-urban divide. Agriculture is the mainstay of most rural areas. It supports roughly two-thirds of the workforce. The agricultural sector has been growing at less than half the pace of the other sectors. During the Seventh Plan, agriculture and allied sectors grew at a rate of 3.4 per cent, while the national economy grew at 6 per cent. In 1997-98, there was a negative growth of 2 per cent in the agricultural sector, although the national economy grew by 5 per cent.

The slower rate of growth of agriculture has serious implications for the rural-urban relationship. In an article in Alternative Economic Survey, Kripa Shankar had shown that this resulted in the further widening of the divide, as the following data (Table 1) relating to agricultural and non-agricultural gross domestic product (GDP) at 1980-81 prices indicate. There had been a further steep rise after the Central government accepted the Structural Adjustment Programme.

Table 1
Agricultural and Non-agricultural GDP Per Worker (at 1980-81 Prices)

Parameter	*1950-51*	*1970-71*	*1990-91*	*1995-96*
Agricultural GDP per worker	Rs. 2,442	Rs. 3,196	Rs. 3,545	Rs. 3,627
Non-agricultural GDP per worker	Rs. 4,469	Rs. 9,179	Rs. 14,660	Rs. 16,715
Ratio	1:1.83	1:2.87	1:4.14	1:4.61

Source: Alternative Economic Survey.

DECLINING NON-FARM EMPLOYMENT

Nearly 91% of employment in the rural areas is in the informal sector. The employment growth rate in the formal sector has steadily declined. Child labour is rampant in India. Under-employment and feminisation of the work force are important consequences of economic reforms.

The introduction of the policy of liberalisation has affected non-farm employment in rural areas. In 1997-98, the annual increase in non-farm employment in rural areas was 4.06 per cent. In 1983-84 it was 3.28 per cent. During 1999-2000 it came down to 2.14 per cent. The consequence has been a very slow reduction in rural poverty. In 1993-94 it was 39.36 per cent, in 1999-2000 the figure came down marginally to 36.35 per cent.

DISPARITIES IN INCOME DISTRIBUTION

Agricultural investments account for 10 per cent of the total investments in the country. The neglect of agriculture and allied sectors is evident from the budgetary allocation. The Human Development Report of India (2002) attempted to divide the rural and urban households on the basis of their incomes as shown in the Table 2.

Table 2
Income Distributions of Rural–Urban Households

Income Groups	*Rural %*	*Urban %*
Low Income (Rs 20000)	65.4	36.7
Lower Middle (Rs 20001-40000)	23.2	33.1
Middle (Rs 40001-62000)	07.5	17.1
Upper Middle (Rs 62001-86000)	02.5	07.8
High income	01.4	05.3

Source: Human Development Report of India (2002).

The NSSO data shows that while 75 per cent of the country's population in 2000-01 resided in rural areas and, they accounted for less than 62 per cent of the total consumption expenditure. The income status is reflected in the per capita consumption expenditure.

INCIDENCE OF POVERTY

If we look at the poverty data (Table 3), a similar situation is noticed. India, a developing economy of over a billion people, recorded a relatively high economic growth during 1980-2000, especially during the 1990s, a decade known for noteworthy structural economic reforms. Development policies focussed on enhanced and targeted public investments in programmes that facilitated improvements in the quality of life of the masses, but the disparity remains.

Table 3
Estimates of Poverty

Year	*All India %*	*Rural %*	*Urban %*
1973-74	54.9	56.4	49.0
1977-78	51.3	53.1	45.2
1983	44.5	45.7	40.8
1987-88	38.9	39.1	38.2
1993-94	36.0 (317)	37.3 (243)	32.4 (75)
1999-2000 (30 day recall)	26.1 (259)	27.1 (195)	23.6 (64)

Note: Figures in parenthesis are population in million

Source: Economic Survey, 2000-2001, Ministry of Finance, Govt. of India

DECLINING INVESTMENT IN SOCIAL SECTOR

According to the estimate of the Eleventh Finance Commission, per capita expenditure on basic services in rural areas during 1997-98 was Rs.24, but in urban areas it was Rs. 49. Rural India contributes 27 per cent to the GDP, but gets back only 5 per cent, which is less than one-fifth of its contribution.

In the Human Development Index prepared by the Planning Commission, there is a significant divide. The value for rural areas is 0.340, in the case of urban areas it is as high as 0.511. The index is a composite of variables capturing attainments in three dimensions of human development namely, economic, educational and health. The same is the situation in respect of the Human Poverty Index: rural 42.25 and urban 44.8. Real spending in the social sector as a percentage of GDP has decreased for the last five years (Table 4). There has been no increase in per capita investment in the social sector.

Table 4
Central Government Expenditure for Social Services

Years	*Budget Estimate as % of GDP*	*Revised Estimate as % of GDP*	*Difference (Rupees in Crores*)*
1996-97	1.20	1.07	1,707
1997-98	1.21	1.08	1,948
1998-99	1.18	1.09	1,601
1999-2000	1.10	1.05	804
2000-01	1.08	–	–

*A Crore is ten million.
Source: Economic Survey 2000-2001.

From the beginning of the economic reforms, the government talked about social safety nets for the poor. A National Renewal Fund (NRF) was planned with significant budgetary allocations to provide social protection for the poor and those adversely affected by the process of liberalization. In practice, the government has done little to strengthen social protection measures. Consistent attempts have been made to undermine budgetary allocations for the social sector in the name of reducing the fiscal deficit. Government expenditure for education, which was 2.75% of GDP in 1998/1999, has declined or remained stagnant. Access to basic health care, especially in the rural areas, remains unavailable to a large majority of people. Central and state government expenditures on health, at 1.11% of GDP in 1998/1999, declined in the last decade. Real spending in the social sector as a percentage of GDP has decreased for the last five years. There has been no increase in per capita investment in the social sector.

CONCLUSION

The year 2001 marks the completion of the first decade of economic reforms. The reform process facilitated the emergence of a new middle class that has appropriated the fruits of economic growth at the expense of the poor and the vulnerable lower middle class. A rupture has developed within the middle class, as a large section of the urban middle class, without adequate technological leverage or bargaining power, is increasingly exposed to unemployment and under-employment.

The economic insecurity of this class of people seems to be contributing to the growing social paranoia and the emergence of reactionary politics in the country. Growing fundamentalism and social tension is clearly linked to growing inequality and marginalisation of a large section of urban population in the liberalised market. The result has been a steady increase of human rights violations, corruption and other criminal activities, social and religious intolerance, and instances of communal violence.

After a decade of liberalization, Government has betrayed the promises of poverty alleviation and social development and created only delusions of development. The government seems to have abdicated and withdrawn from its responsibilities. The ten years of reform facilitated the withering away of the welfare state and left the vast majority of Indians, especially in the rural areas, with shattered dreams of sustainable social and economic development.

2

RURAL URBAN DISPARITIES– GUJARAT VIS-À-VIS INDIA

Rural and urban disparity is studied in terms of demographic characteristics, economic dimensions, availability of social and physical infrastructure to improve quality of life, consumptions patterns and incidence of poverty. Under the demographic characteristics, we have considered rural-urban variations in population, sex ratio and population density, people below poverty line, vital rates, namely, birth rate, death rate, and infant and child mortality rate.

DISPARITY ACROSS DEMOGRAPHIC DIMENSIONS

Gujarat is one of the developed states in India. It has progressed to achieve the fourth rank in per capita income among major states in India. Today it is one of the prosperous states of India with about 50 million population in 2001 spread over 196,000 sq. km. Though the state has less than 5 per cent of population, it contributes about 7 per cent of the production and about 13 per cent of the industrial output.

Population and Density

Population of Gujarat (Table 1) increased from 9.1 million in 1901 to 50.05 million in 2001.

Table 1
Trends in Population and Density

Year	Population (millions)		Sex ratio		Inter-census Growth rate p. a. (%)		Population density per sq. km.		Gujarat				India			
									Rural		Urban		Rural		Urban	
	Guj	India	Guj	India	Guj	India	Guj	India	CBR	CDR	CBR	CDR	CBR	CDR	CBR	CDR
1901	9.1	238.3	954	972	–	–	46	77	–	–	–	–	–	–	–	–
1911	10.0	252.0	946	964	0.8	0.6	50	82	–	–	–	–	–	–	–	–
1921	10.2	251.2	944	955	0.4	0.0	52	81	–	–	–	–	–	–	–	–
1931	11.5	278.9	945	950	1.2	1.1	59	90	–	–	–	–	–	–	–	–
1941	14.5	318.5	941	945	1.8	1.3	70	103	–	–	–	–	–	–	–	–
1951	16.3	361.0	952	946	1.7	1.3	83	117	–	–	–	–	–	–	–	–
1961	21.0	439.1	940	941	2.4	2.0	105	142	–	–	–	–	–	–	–	–
1971	27.0	548.2	934	930	2.6	2.2	136	178	42.1	18.1	36.1	13.1	38.9	16.4	30.1	9.7
1981	34.1	683.3	942	933	2.4	2.2	174	221	36.1	12.4	29.8	10.7	35.6	13.7	27.0	7.8
1991	41.3	46.8	934	927	1.9	2.1	211	273	28.2	8.8	25.9	7.9	30.9	10.6	24.3	7.1
2001	50.05	102.7	921	927	2.05	2.1	258	324	26.9	8.8	21.5	5.6	27.1	9.0	20.2	6.3

Note: Growth rate for 1961-71 and 1971-81 were calculated taking into the fact that the reference date of 1971 census was April, whereas 1961 and 1981 censuses had the usual reference date of March 1.

Source: Based on respective population censuses. [From Table 1 (Hirway and Mahadevia, 2004: 22)]

The compound annual rate of growth (CARG) during the 1950s, 1960s and 1970s was fairly high, but declined considerably between 1971 and 1991, from 2.6 per cent between 1961-71 to 2.4 per cent between 1971 - 1981 and to 1.9 per cent between 1981-1991 (Hirway and Mahadevia, 2004: 21).

The main reason for the rapid decline in population in the state since 1971 has been the decline in CBR (crude birth rate); which declined from 42.1 in 1971 to 36.1 in 1981 to 28.2 in 1991 in rural areas (indicating 14 percentage point decline in two decades) and from 36.1 in 1971 to 29.8 in 1981 to 25.9 in 1991 in urban areas (more than 10 percentage point decline in two decades). In the 1990s, a deceleration in the decline in the CBR in the state was observed. The rural CBR fell only marginally, by 2 per cent points, while the urban CBR fell by 4 per cent points (Hirway and Mahadevia, 2004:22). The sex ratio in the state has continuously declined from 942 in 1981 to 934 in 1991 and to 921 in 2001. The rate of decline accelerated during 1991-2001, from 8 points during 1981-91 to 13 points during 1991-01. The crude birth rate (CBR) in Gujarat was 25.2 in 1999-00 and 24.9 in 2001 (Hirway and Mahadevia, 2004: 23).

According to 2001 census (Table 2), 72.15 % of the 1027 millions of the Indian population live in rural areas. Whereas in Gujarat 62.65 % of the 50.6 millions of Gujarat population live in rural areas making urbanization in Gujarat well above the urbanization at all India level. Among the States, Gujarat is the fourth most urbanized with proportion of urban population at 34.5 per cent state compared to 25.7 per cent for the country according to 1991 census. Urbanization in Gujarat reached 37.35 per cent in 2001 and is projected to reach 42.2 per cent in 2011.

Table 2
Rural-Urban Population[1] (2001)

(In Millions)

		Rural[2]			*Urban*		
State	*Male*	*Female*	*Total*	*Male*	*Female*	*Total*	*Total*
India	381	360	741	150	135	285	1027
%	51.42	48.58	72.15	52.63	47.37	27.75	100
Gujarat	16.29	15.41	31.7	10.1	8.87	18.9	50.6
%	51.39	48.61	62.65	53.24	46.76	37.35	100

Note: The population of India and the estimated population of entire Kutch district, Morive, Mallkiya-Miyana and Wankaner talukas of Rajkot district, Jodiya taluka of Jamnagar District of Gujarat State and entire Kinnaur district of Himachal Pradesh were population enumeration of Census of India, 2001 could not be conducted due to natural calamity.

It may be noted that sex ratio (Table 3) in terms of females for 1000 males has shown some marginal

Table 3
Sex Ratio and Population Density[3]

	Sex Ratio (Females per 1000 Males)				Population Density (Per Sq. km.)			
	1991[4]			2001	1991			2001
State	*Total*	*Rural*	*Urban*	*Total*	*Total*	*Rural*	*Urban*	*Total*
India	927	939	894	933	267	214*	3,370*	324
Gujarat	934	949	907	921	211	142	2,773	258

* Excludes the figures for Arunachal Pradesh and Sikkim for which the area figures for rural and urban for these states are not available.

Note: 1. The population of India and the estimated population of entire Kutch district, Morive, Mallkiya-Miyana and Wankaner talukas of Rajkot district, Jodiya taluka of Jamnagar District of Gujarat State and entire Kinnaur district of Himachal Pradesh were population enumeration of Census of India, 2001 could not be conducted due to natural calamity.

2. For working out the population density of India and Jammu and Kashmir, the entire area and population of those areas of Jammu and Kashmir which are under illegal occupation of Pakistan and China have not been taken into account.

improvement at all India level whereas the sex ratio is declining in Gujarat. Change in the population density is of same order (22% in case of Gujarat and 21% in case of All India). When compared to the density of population at all India level, population density in Gujarat is less, largely due to sparsely populated and inhospitable regions in Kutch.

Vital Rates and Health Indicators

There has been a decline in the birth, death and infant mortality rate (Table 4). The death and infant mortality rate in Gujarat is less than the all India. But the birth rate in Gujarat is higher than all India.

Table 4
Vital Rates[4]

	Crude Birth Rate		*Crude Death Rate*		*Infant Mortality Rate*	
State	*1991*	*2001*	*1991*	*2001*	*1991*	*2001*
India	29.5	24.8	9.5	8.9	80	66
Gujarat	27.5	25.4*	8.5*	7.9	69	60

* Year 2000-2001.

Table 5
Vital Rates in Rural Areas[5]

(Per thousand)

	Birth Rate		*Death Rate*		*Infant Mortality Rate*	
State	*2000*	*2001*	*2000*	*2001*	*2000*	*2001*
Rural India	27.6	27.1	9.3	9.0	74	72
Rural Gujarat	26.8	26.6	8.3	8.8	69	67

Relative prosperity of Gujarat extends to rural areas as well and the same is reflected in better birth, death and infant mortality rates in comparison with Rural India (Table 5).

During the period 1990 to 1997 (Table 6), the decrease in infant mortality rate in rural areas (-10.45 %) is little higher than that of the urban areas (-10 %). Also the decrease in post neo-natal mortality rate is higher in rural areas when compared to the urban areas. But the rate of decrease of Neonatal mortality rate and Perinatal mortality rate is higher in case of urban areas than in the rural areas. However, it is a matter of a concern still that infant and child mortality rates are considerably high when compared to those in the urban areas.

Table 6
Infant and Child Mortality Rates- All India 1990-97

	Infant Mortality Rate			*Neo-natal Mortality*			*Post-neo Natal Mortality*			*Peri-natal Mortality*		
Years	*R**	*U**	*T**	*R*	*U*	*T*	*R*	*U*	*T*	*R*	*U*	*T*
1990	86	50	80	57.4	30.9	52.5	28.9	19.5	27.2	51.7	34.0	48.4
1991	87	53	80	55.4	32.2	51.1	31.3	21.6	29.3	49.1	32.7	46.0
1992	85	53	79	53.6	33.0	50.0	31.4	21.0	29.4	50.3	34.0	47.5
1993	82	45	74	52.3	28.4	47.1	28.2	15.8	25.2	47.9	31.0	44.2
1994	80	52	74	52.0	32.6	47.7	27.8	19.6	26.0	43.4	38.7	42.5
1995	80	48	74	52.3	29.2	47.1	27.8	19.0	25.9	17.6	31.2	44.6
1996	77	46	72	50.0	28.0	47.0	27.0	17.0	25.0	46.0	32.0	44.0
1997	77	45	71	50.5	26.2	46.1	26.5	19.0	25.1	46.3	29.0	43.2

* R = Rural, U = Urban, T = Total

DISPARITY ACROSS ECONOMIC DIMENSIONS

Net domestic product per capita, rural urban consumption divide, employment and work force distribution and incidence of poverty are considered.

Net Domestic Product

The Net Domestic Product per capita at current prices is presented in Table 7. However the large sections of population continue to live in rural areas lower net domestic product per capita in rural areas mean lower labour factor productivity.

Table 7
Net Domestic Product Per Capita at Current Prices6- All India

	1970-71			*1980-81*			*1993-94*		
	Rural	Urban	Total	Rural	Urban	Total	Rural	Urban	Total
NDP Per Capita	529	1294	680	1245	2888	1625	5876	14035	8037
Rural as % of Urban		40.88	77.79		43.11	76.62		41.87	73.11

Persistent Rural-Urban Consumption Divide

Urban-rural per capita consumption expenditure ratios are presented in Table 8. It can be seen that at all India level urban and rural disparity in per capita consumption has shown slight decline in late seventies but gradually increased thereafter. In relative terms urban and rural disparity in per capita consumption expenditure in Gujarat was much less compared to all India and there is a sharp increase since mid 19990s.

Data from the National Sample Survey on household consumer expenditure in India are useful not only in evaluating living standards across the country but also in measuring income inequalities. The report of the 60th round was released on November 23, 2005. Although these data are primarily viewed for what they say on living standards, they have one other focused use: changing consumption patterns can be interpreted for their insights into income differentials, despite the fact that disparities in consumption need not always correlate to income inequalities. As a rule, the proportion of consumption expenditure decreases when income rises. The rich save more and consume less in relative terms. The NSS consumer data are segmented into rural and urban areas on one hand and among States and geographical regions on the other. The latest survey covering the period Jan-June 2004 captures certain well-known trends that previous surveys revealed (*Source: URL:http://www.thehindu.com/thehindu/biz/2005/12/19*).

Table 8
Urban-Rural Monthly Per Capita Consumption Expenditure Ratios[7]

(In Current prices)

State	*1972-73*	*1977-78*	*1983*	*1987-88*	*1993-94*	*1999-00*	*2004-05*
India	1.43	1.39	1.46	1.69	1.62	1.87	1.88
Gujarat	1.11	1.44	1.33	1.44	1.46	1.68	na

(1) Urban households incur almost twice the expenditure incurred by rural households. During this period, an average urban household had an MPCE (monthly per capita consumption expenditure) of Rs. 1,060, which was 88 per cent more than the MPCE of an average rural household (Rs. 565). The difference between rural and urban expenditure has remained constant over various periods covered by NSS.

(2) The share of food in the average consumption basket has shown a steady decline in both urban and rural areas. While income levels have been rising everywhere, urban households have been spending less on food in proportionate terms compared to rural households. Almost 54 per cent of the average rural MPCE was on food. Urban households on the other hand spent almost 58 per cent on items other than food, with fuel and lighting, clothing and footwear accounting for a big chunk.

(3) Again, within the food basket, the share of cereals has been coming down. In rural areas, cereals accounted for 18 per cent of MPCE, a sharp fall from the 26 per cent recorded in 1987-88. In urban areas, it was 10 per cent in 2004-05, down from 15 per cent earlier. The sharp disparity between rural and urban areas extends to education (rural 3 per cent, urban 7 per cent) and miscellaneous consumer services (rural 3 per cent, urban 7 per cent).

(4) Across States, there are significant disparities in consumption expenditure. Jharkhand, Orissa, Chattisgarh, Madhya Pradesh and Bihar had an MPCE less than Rs. 450 in rural areas. Some of these States have fared poorly in terms of urban MPCE as well. On the other side, Kerala topped the list under both urban and rural MPCE. (Rs. 990 rural and Rs. 1,372 urban).

(5) The use of kerosene as cooking fuel has been declining sharply in both rural and urban areas. However, around 46 per cent of rural households depend on kerosene for lighting purposes. Until rural electrification gathers pace, the dependence on kerosene will remain. What specific issues do these data address?

Trends in Unemployment

There was a small decline in unemployment both in urban and rural areas up to early 1990s and there after unemployment has increased more so in the rural areas as can be seen from the Table 9. It is disconcerting to note that India seems to have entered a phase of "jobless" growth and development beginning in 1990s.

Table 9
Trends in Unemployment in India[8]

(In Million)

Particulars	*1983*	*1993*	*1999-2000*
Total unemployed	21.61	20.24	26.44
Total rural unemployed	16.81	14.30	19.31
Total urban unemployed	5.43	5.94	7.12
Percentage of rural unemployed to total	74.85	70.65	73.05

Growth rate of unemployment is -0.62 % between 1993 and 1983 largely due to reduction in rural unemployment not withstanding the marginal increase in urban unemployment. It appears after 1993, there has been a sharp increase both in rural (from-1.16 during 1983-1993 to 5.13 during 1993-2000) and urban (from 0.85 during 1983-1993 to 3.07 during 1993-2000) unemployment with the sharper increase rural unemployment.[9]

Rural Workers' Distribution

Rural work participation rates, both male and female as well as in over all terms are higher for Gujarat when compared with All India (Table 10). While over all work participation rate in rural Gujarat is higher by about 5 %; it is higher by about 3 % in case of male and 8% in case of male. This higher female work participation rate is higher–significantly not in cultivators and Agri. Labour categories; less in HH industry category–but in other workers category. This is in line with the general observation of positive correlation between urbanization and female work participation. The distribution of rural workers vis-à-vis all India in the year 2001[10] is mentioned hereunder:

Table 10
Distributions of Rural Workers (2001)

Occupation	*State*	*Total*	*Male*	*Female*
Cultivators	India	40.1 %	42.2 %	36.5 %
	Gujarat	38.3 %	43.0 %	31.2 %
Agri. Labourers	India	33.2 %	27.5 %	43.4 %
	Gujarat	33.6 %	26.7 %	44.0 %
HH Industry	India	3.8 %	2.8 %	5.4 %
	Gujarat	1.6 %	1.6 %	1.4 %
Other workers	India	22.9 %	27.5 %	14.7 %
	Gujarat	26.6 %	28.7 %	23.4 %
Total workers	India	42.0 %	52.4 %	31.0 %
	Gujarat	47.6 %	55.6 %	39.0 %

Incidence of Poverty

According to figures available for 1999-2000 (Table 11), incidence of poverty in Gujarat, both in rural as well as urban areas is much less. Closer examination of data in the table below reveals that incidence of poverty in rural Gujarat is less by about 14%; in case of urban areas it is less by about 8 % and in over all terms it is less by about 12%. This makes rural Gujarat a relatively much prosperous than much of the rural India.

Table 11
Populations below Poverty Line[11] (1999-2000)

Sector	*Particulars*	*India*	*Gujarat*
Rural	No. of persons (Lakhs)	1932.43	39.80
	% of persons	27.09	13.17
	Poverty line (Rs.)	327.56	318.94
Urban	No. of persons (Lakhs)	670.07	28.09
	% of persons	23.62	15.59
	Poverty line (Rs.)	454.11	474.41
Combined	No. of persons (Lakhs)	2602.50	67.89
	% of persons	26.10	14.07

DISPARITY IN QUALITY OF LIFE

There is considerable disparity in rural and urban areas with regard to availability of social infrastructure like education and health facilities and the quality, adequacy and actual availability of services. Rural people spend relatively more to access quality services in education and health. Education–literacy, educational levels, education level by gender, education status in rural and urban areas, education of children, availability of primary health care facilities, vaccination services were considered. Physical infrastructure in terms of access, housing, lighting, electricity, cooking fuel, drinking water, sanitation, public distribution system, and public communication facilities was also examined.

Education-Literacy

There is a gradual increase in the literacy rate[12] (Table 12) among the urban population from 73.10 % in 1991 to 80.30 % in 2001 when compared to a substantial increase in the literacy rate among the rural population of country from 44.70 % to 59.40 % in 2001.

Educational Levels in Rural and Urban Areas

The percentage of population who are literate and up to Primary is higher in case of rural population (19.4 %) when compared to urban population (16.1 %). But the percentage of literate population under Higher Secondary, Graduate and above is substantially more in urban areas than in rural areas (Table 13).

Educational Status of Rural Population

The literacy rate in the State (excluding children in the age group 0-6 years) has increased from 61.29 per cent in 1991 to 69.14 per cent in 2001.The rate of increase of rural literacy rate in Gujarat state is less when compared to all India. However in Gujarat, the increase in the literacy

Table 12
Literacy Rates-All India[13]

(In percentage)

Year	*Sector*	*Male*	*Female*	*Persons*
1951	Rural	19.02*	4.87*	12.10*
	Urban	45.06*	22.33*	34.59*
	Total	27.16	8.86	18.33
1961	Rural	34.30	10.10	22.50
	Urban	66.00	40.50	54.40
	Total	40.40	15.35	28.30
1971	Rural	48.60	15.50	27.90
	Urban	69.80	48.80	60.20
	Total	45.96	21.97	34.45
1981	Rural	49.60	21.70	36.00
	Urban	76.70	56.30	67.20
	Total	56.38	29.76	43.57
1991	Rural	57.90	30.60	44.70
	Urban	81.10	64.00	73.10
	Total	64.13	39.29	52.21
2001	Rural	71.40	46.70	59.40
	Urban	86.70	73.20	80.30
	Total	75.85	54.16	65.68

Note : (*) for 1951 the population male, female and persons refers to effective literacy rates and the break up of rural, urban and male-female components are crude literacy rates.

Table 13
Educational Levels in Rural and Urban Areas[14]

Education	*1993–1994*				*1999–2000 (Age 7 and above) %*	
	Rural (%)		*Urban (%)*			
	Male	*Female*	*Male*	*Female*	*Rural*	*Urban*
Not literate	45.50	67.90	24.10	38.40	44.0	20.2
Literate					56.0	79.8
Literate & up toPrimary	33.70	23.00	33.20	31.00	19.4	16.1
Primary					13.5	14.7
Middle	10.90	5.60	14.50	11.80	12.3	16.5
Secondary & above	9.80	3.40	28.10	18.70	6.4	13.5
Higher Secondary					2.8	8.5
Graduate & above					1.7	10.5

rate among rural women (from 28.8 % in 1981 to 45.57 % in 2001) is comparatively more than the increase in literacy rate among rural men (from 57.76 % in 1981 to 70.71 % in 2001). Gender wise literacy figures are presented below in Table 14.

Table 14
Literacy Rates in Rural Areas

(In percentage)

Literacy	*1981* Male	*1981* Female	*1981* Total	*1991* Male	*1991* Female	*1991* Total	*2001* Male	*2001* Female	*2001* Total
India	49.59	21.70	36.01	57.87	30.62	44.69	71.18	46.58	59.21
Gujarat	57.76	28.80	43.57	66.84	38.65	53.09	70.71	45.75	58.53

Note: Literacy rates relate to the persons of the age group of 7 years and above.

The percentage of literate population in all educational level in rural areas of Gujarat is more than that of all India (Table 15).

Table 15
Education Status of Rural Population[15]

General Education Level	*Male Persons* India	*Male Persons* Gujarat	*Female Persons* India	*Female Persons* Gujarat
Not Literate	41.2	34.6	61.5	57
Literate & up to Primary	34.2	36.4	26.0	28.4
Middle	12.6	16.1	7.5	8.8
Secondary & above	11.7	12.9	5.0	5.8

Rural Health Services

Vaccination among Children in Rural India

Except for Polio O, the percentage of children vaccinated is higher in Gujarat when compared to all India (Table 16), may be due to higher number of child births taking place outside health centers. The percentage of children with vaccination card in Gujarat is marginally less than that of all India.

Table 16
Vaccinations among Children in Rural India (1998-99)[16]

(Percentage)

Percentage Vaccinated	*India*	*Gujarat*
BCG	71.6	84.7
Polio O	13.1	5.3
DPT 1	71.4	83.1
DPT 2	65	75.4
DPT 3	55.1	64.1
Polio 1	83.6	90.2
Polio 2	78.2	82.5
Polio 3	62.8	68.6
Measles	50.7	63.6
All	42	53
None	14.4	6.6
% showing vaccination card	33.7	31.8

Note: Table includes only surviving children from among the two most recent births in the three years preceding the survey. BCG, Measles and three doses each of DPT and Polio vaccines (excluding Polio O).

Health Care Facilities in Rural Areas

It appears that Gujarat has better provisioning of health care facilities through sub centers in terms of population norms and need to improve the ratio in case of PHCs and CHCs (Table 17).

Table 17
Health Care Facilities in Rural Areas-1999[17]

State	*Sub Center Health*	*Primary Health Centers*	*Community Centers*
National Population Norms[18]	*3000-5000 Pop.*	*20000-30000 Pop.*	*About 1 Lakh*
India (Actual)	134108	22349	5587
India (Ideal as per norms)	About 158000	About 25000	About 6300
Shortfall	−15.12 %	−10.60 %	−11.32 %
Gujarat (Actual)	7247	1044	253
Gujarat (Ideal as per norms)	About 6800	About 1100	About 270
Shortfall	+ 6.57	- 5.09	- 6.30

Note: Rural population of India and Gujarat as per 1991 Census is 62.86 & 2.71 Crores.

Rural Housing

The percentage of people residing in Pucca and Semi-pucca house is more in Gujarat when compared to all India (Table 18). Also the percentage of people residing in Kuccha house is also less when compared to all India.

However during the period 1981–1994 in Gujarat, there is marginal increase in population residing in Pucca house, a decline in the population residing in Semi pucca house and an increase in percentage of people residing in Kuccha house. Where as in all India, during the same period there is decline in population residing in Pucca house and Kuccha house and marginal increase in population residing in Semi pucca. Estimated rural housing shortage in 1991 was about 265000 and is 5.52 %.

Table 18
Rural Housing[19]

Rural Housing	*1981*	*1991*	*1993–1994 NSS 50th Round*
*All India**	%	%	%
Pucca	22.53	30.59	29.2
Semi-pucca	36.93	35.65	38.10
Kuccha	40.55	33.76	32.70
Gujarat			
Pucca	36.42	43.42	36.70
Semi-pucca	52.66	51.61	44.20
Kuccha	10.91	04.97	19.10

*Excludes data for Assam in 1981 and Jammu and Kashmir in 1991.

Lighting in Rural Households

5.02 lakh villages out of 5.79 lakh revenue villages (1981 Census) of the country were electrified by March 1998. In Gujarat, 17936 villages out of 18114 revenue villages (1981 Census) were electrified. The percentage of households[20] with electricity is higher in both rural and urban areas in Gujarat when compared to all India (Table 19). Also the increase in percentage of households with electricity over the period 1991 to 1994 is higher in Gujarat when compared to all India.

Table 19
Lighting in Rural Households[21]

	1991			*NSS 50th Round (July 1993–June 1994)*	
State	*Rural %*	*Urban %*	*Total %*	*Rural %*	*Urban %*
All India	30.54	75.78	42.37	37.1	82.8
Gujarat	56.43	82.96	65.93	67.5	89.7

Percentage of rural households using Electricity as source of lighting is higher in Gujarat when compared to all India (Table 20). Whereas the percentage of rural households using Kerosene and Solar energy is lower in Gujarat when compared to all India. The percentage of rural households having no lighting facility is higher in Gujarat when compared to all India.

Table 20
Primary Source of Lighting (2001)[22]

Source of lighting	*India %*	*Gujarat %*
Electricity	43.52	72.12
Kerosene	55.61	26.17
Solar energy	0.29	0.21
Other oil	0.11	0.20
Any other	0.16	0.31
No lighting	0.31	0.99

Fuel for Cooking in Rural Households

Both in rural India as well as in Gujarat, fuel wood is used as the primary source for cooking in rural households (Table 21). Percentage of rural households using crop residue, cow dung, coal, lignite, charcoal and electricity is less in Gujarat when compared to all India. The percentage of rural households using LPG, Kerosene and Fuel wood is higher in case of Gujarat when compared to all India. Also the percentage of people having no cooking facility is higher in Gujarat when compared to all India.

Table 21
Fuel for Cooking in Rural Households (2001)[23]

Fuel Used for Cooking	*India %*	*Gujarat %*
Firewood	64.10	65.08
Crop Residue	13.10	12.86
Cowdung Cake	12.80	5.33
Coal, Lignite, Charcoal	1.07	0.09
Kerosene	1.62	5.98
LPG	5.67	9.17
Electricity	0.13	0.02
Biogas	0.47	1.07
Any other	0.82	0.07
No Cooking	0.22	0.32

Drinking Water, Electricity and Toilet Facility in Rural Households

Almost half of the rural households in Gujarat have tap water facility and also the percentage of rural households with tap water facility is much higher when compared to all India (Table 22). Also the percentage of rural households dependent on hand pump, tube well, well, tank, pond, river, canal and spring for drinking water is less in Gujarat when compared to all India.

Percentage of rural households with water, electricity as well as toilet facility is higher in Gujarat when compared to all India is presented in Table 23.

Rural Coverage of PDS

PDS users among rural areas are higher in Gujarat when compared to all India (Table 24). But in urban areas the

PDS users in Gujarat is lower when compared to all India. The rural PDS users among all users, PDS users among non-poor population are higher in Gujarat when compared to all India. The non-poor among all users are lower in Gujarat when compared to all India.

Table 22
Rural Households by Source of Drinking Water (2001)[24]

Source of Drinking Water	*India %*	*Gujarat %*
Tap	24.29	49.09
Hand Pump	43.20	22.76
Tubewell	5.74	5.02
Well	22.23	18.27
Tank, Pond, Lake	1.28	0.55
River, Canal	1.33	0.69
Spring	0.93	0.29
Any other	1.01	3.33

Table 23
Water, Electricity and Toilet Facilities in Rural Households (1991)[25]

Facility	*India %*	*Gujarat %*
Water	55.92	60.04
Electricity	31.10	56.43
Toilet facility	8.84	11.16
All three facilities	Not available	8.79

*Excludes Assam and Jammu & Kashmir.

Table 24
Rural Coverage of PDS[26]

Rural Coverage of PDS (1986-87)	*India (In %)*	*Gujarat (In %)*
PDS Users Among all Persons in Rural Area	25.12	44.60
PDS Users Among all Persons in Urban Area	35.74	31.98
Rural PDS Users to all Users	71.30	76.70
PDS Users Among Poor Population	31.00	54.40
PDS Users Among Non Poor Population	26.33	36.73
Non Poor Among All Users	72.53	68.95

Communication

There are more number of post offices to cater to the rural population in Gujarat when compared to the population of the country (Table 25).

Table 25
Rural Post Offices (March 1994)[27]

State	Number	Rural Population Per P.O.	Area Served by a P.O. (Sq. KM)
India	135982	5740	21.49
Gujarat	8085	4777	22.01

Notes

1. Source: G O I, Census of India, 2001, Provisional Population Totals, Paper–I of 2001, Registrar General, India.
2. Source: G O I, Census of India, 2001, Provisional Population Totals, Paper–I of 2001, Registrar General, India.
3. Source: Financial Express 29-03-2001, P.7, Col. No. 1-7.
4. http://www.censusindia.net/ # Source: http://cbhidghs.nic.in/KeyHealthInd93.asp, http://www.bisnetworld.net/bisnet/states/gujrat1a.htm, http://www.azimpremjifoundation.org/html/statistics.htm
5. Source: G O I, Sample Registration System (SRS), October 2002, Registrar General, India.
6. Source: Rural Development Statistics 2002-2003, NIRD (p. 39).
7. Approach Paper on growing rural-urban disparity, pp. 8-9.

8&9. Source: Computed on the basis of NSS reports-various issues 1. Report on the third Quinquennial Survey on Employment and Unemployment: Survey results-All India Reports No. 341; 2. Report on the third Quinquennial Survey on Employment and Unemployment: Survey results– All India Reports No. 341/1 to 341/15; 3. Employment and

Unemployment in India 1993-94, Fifth Quinquennial Survey, NSS 50th round, report Number 409 and 4. Employment and Unemployment situation in India, NSS 55th Round, Report No. 458

10. Source: GOI, Provisional Population Totals, Paper-3 of 2001, Distribution of Workers and Non-workers, Series 4, Punjab, Census of India 2001.
11. Source: Planning Commission, National Human Development Report, 2001.
12. Source: Planning Commission, National Human Development Report, 2001.
13. Source: Ministry of Finance, Economic Survey 2001-2002.
14. Source: NSSO Sarvekshana Vol. XX, No. 1 68th issue and NSSO Sarvekshana Vol. XXIV, No. 2 & 3, 85th issue October 2000-March 2001, Journal of NSSO.
15. Source: NSSO, Employment & Unemployment situation in India, Part-I, 1999-2000, 55th Round (July 1999-June 2000).
16. *Source:* International Institute of Population Sciences, National Family Health Survey (NFHS-2), India, 1998-99.
17. Source: Planning Commission, Annual Plan 2002-03.
18. Rural Development Statistics, 2002-03, pp. 264.
19. Source: C S O, Statistical Abstract India, 1998 & CMIE, India Social Sector, February 1996.
20. Source: C S O, Registrar General, Census of India 1991.
21. Source: CMIE, Energy, March-April, 1999 & Economic and Political Weekly, January 13, 2001.
22. Source: GOI, Census of India 2001, Tables on Houses, Household Amenities and Assets, India.
23. Source: GOI, Census of India 2001, Tables on Houses, Household Amenities and Assets, India.

24. Source: GOI, Census of India 2001, Tables of Houses, Households Amenities and Assets, India.
25. Source: GOI. Census of India 1991, Series-I, Paper-2 of 1993, Registrar General of India.
26. Source: Agriculture Situation in India, July 1995, pg. 205.
27. Source: Department of posts, Annual report 1994-95.

3

RURAL URBAN DISPARITY IN GUJARAT

The enterprising population of Gujarat, supported by its `progressive' leadership, has brought the state to the forefront of economic development in the country. The state has not only acquired and maintained its fourth rank in per capita NSDP, but has also made a quantum jump in the post-liberalisation period by attracting almost all the highest industrial investments, particularly in the large and medium industry sector, and has consequently experienced the highest growth in per capita Net State Domestic Product during this period (Hirway and Mahadevia 1999). The average annual rate of growth of the state economy (at the 1980-81 prices) in the post-reform period, i.e. from 1990-91 to 1997-98, has been 8.65 per cent, which is among the highest in India. In terms of the growth rate of the per capita state domestic product too, the state is at the top with more than 6 per cent of annual rate of increase.

Structure of the Gujarat economy, sectoral growth rates, income, labour productivity were examined under the Gujarat state's macroeconomic features available workforce and its employment across industrial categories, across sectors and by type and work participation rates were examined. Incidence of poverty rural and urban areas and across regions and sectors was also examined. Quality of life as reflected in the life expectancy at birth, infant mortality rate and expenditure on health were considered.

FEATURES OF GUJARAT ECONOMY

The share of Gujarat state at current prices and at constant (1993-94) prices in Gross Domestic Product at All India level works out to 6.15 per cent and 6.49 per cent respectively. The per capita income at constant prices has been estimated at Rs. 13715 in 2002-03, higher than National average of Rs. 10964; as against Rs. 13684 in 2001-02, registering an increase of 0.2 per cent over the previous year whereas the per capita income at current prices has been estimated at Rs. 22047 in 2002-03 higher than the National average of Rs. 18912; as against Rs. 20695 in 2001-02, registering an increase of 6.5 per cent over the previous year.

Gujarat has highly diversified sources of income: about 78 per cent of the State Domestic Product in Gujarat comes from non-primary sectors, as against 60 per cent of the national economy. Similarly, Gujarat has a relatively diversified workforce, with more than 40 per cent employed in non-primary sectors, as against 32.6 per cent of the all-India workforce. The money and capital markets in the state are well developed, allowing for a relatively easy access to finance for its enterprising population. It appears that the state has done relatively well for itself during the last three to four decades.

The economy grew at 5.2 per cent in the 1980s and this was largely due to the growth in the secondary sector, which is the backbone of Gujarats' economy (Rani, 2004). The growth rate of 7.2 per cent in the 1990s was due to emphasis towards infrastructure development leading to industrialization (Table 1).

Table 1
Sectoral Growth Rate in Income at 1980-81 Prices[1]

(In Percentage)

Sectors	*1960-61 to1969-70*	*1970-71 to1989-90*	*1980-81 to1989-80*	*1990-91 to1996-97*
Primary	0.81	2.90	1.39	3.87
Secondary	4.86	5.85	7.28	10.08
Tertiary	3.25	5.66	6.60	6.17
All Sectors	2.52	4.36	5.21	7.18

This approach helped the state to attract the highest industrial investment in the large and medium sector – more than 17000 billion during 1991-97 (Hirway 2000:3110). The major development was the restructuring of the industrial sector towards chemical based industries in the Vadodara–Valsad belt that is called the Golden Corridor. There was also an overall improvement in physical infrastructure, especially in Saurashtra-Kutch region where clusters had come in the late 1980s. This includes the ship-breaking yard near Bhavnagar, the diamond industry in Amreli and Bhavnagar and mineral based industries in coastal Saurashtra, in addition to the already existing developed clusters of engineering industry in Rajkot. Also there was a boom in the construction industry, especially in the urban areas. It can be concluded that Gujarat is following a development pattern of developed economies where the economy becomes highly industrialized before the service sector expands.

The critical question, however, is: how has this development treated the poor, and what are the future prospects of the economy as far as the poor are concerned? On the face of it, it seems that the incidence of poverty is still quite high in the state and the prospects for the poor at the bottom of the economy are not very bright. How can this happen in a state which is at the forefront of economic development in the country and which is growing at a rapid rate at present?

Sectoral Growth Rates

Gujarat's economy has shown a consistently increasing rate of growth during the past few decades (Table 2). CARG has gone up from 3.32 per cent in the 1960s to 4.88 per cent in the 1980s, and 5.53 per cent in the 1990s. If the primary sector is excluded, the state economy shows about 7.7 per cent growth during the past two decades, which is indeed quite high. Per capita income in the state has increased at the rate of 3.14 per cent in the 1980s and 4 per cent in the 1990s (Hirway and Mahadevia, 2004: 24).

Table 2
Sectoral Growth Rates Based on Constant Price (1980-81) in Gujarat (%)

Particulars	*1960-61 to1970-71*	*1970-71 to1980-81*	*1980-81 to1990-91*	*1990-91 to1997-98*	*1980-81 to1997-98*
Primary Sector	2.91	4.15	-1.15	3.46	1.42
(Agriculture) %	(2.27)	(4.22)	(-0.81)	(3.91)	(1.34)
Secondary Sector	3.62	5.64	7.89	12.83	8.87
(Manufacturing)%	(3.04)	(5.55)	(8.59)	(13.99)	(9.19)
Tertiary Sector %	3.51	5.86	7.44	8.73	7.25
Overall SDP %	3.32	4.95	4.88	8.99	6.00
India GDP %	3.17	3.66	5.42	5.07	5.44

Source: EPW Research Foundation (1998), Directorate of Economics and Statistics 2002, Gandhinagar

The sectoral growth rates in the state economy appear to be fairly balanced during the 1960s and the 1970s. The rates of growth of the primary secondary and tertiary sectors were respectively 2.91, 3.62 and 3.51 in the 1960s and 4.15, 5.64 and 5.86 in the 1970s. This assured close linkages between agriculture and industries during these decades.

Since the 1980s, however, one observes that the primary sector, particularly agriculture, has tended to lag far behind. The annual growth rate in agriculture became negative in the 1980s when the secondary and tertiary sectors registered a big jump to 7.85 per cent (8.73 per cent in manufacturing) and 6.83 per cent respectively.

The situation did not improve even in the 1990s, with the long-term rate of growth of the primary sector remaining below 1 per cent during this period. Considering the fact that around 60 per cent of the state population still depends on agriculture, the lagging behind of this sector clearly implies low benefits to the majority of the population.

Not only has agriculture lagged behind, it has also proved to be highly unstable. This is reflected in the high fluctuations in the per capita State Domestic Product from the primary sector. While the per capita incomes in the secondary and tertiary sectors show a consistent increase during the last 15-20 years, those in the primary sectors fluctuate widely without showing any long-term trend. In short, the seemingly high structural transformation of the Gujarat economy is in fact not all that healthy.

This pattern of growth is not likely to sustain for long as it does not allow agriculture and industry to develop strong and mutually reinforcing linkages, which are so very essential for sustained development.

Per Capita NSDP in Different Sectors in Gujarat

The Gujarat State is experiencing a rapid industrial and economic growth and can be considered amongst the best in the country. The economy of Gujarat grew at the rate of 4.93 percent during the decade 1981-91 and marked a growth rate of about 7.7 percent between 1993-94 and 1998-99. The per capita income at constant (1993-94) prices is estimated at Rs. 13,709/- for 1998-99.

A striking feature of sectoral incomes in Gujarat is the highly fluctuating character of per capita income from the primary sector, which varies from Rs. 792 in 1980-81 to Rs. 729 in 1990-91 to Rs. 896 in 1994-95 and to Rs. 491 in 2000-2001 (Table 3).

Over a period of two decades i.e. from 1981 to 2001, the per capita income has decreased by 38 percent in case of primary sector where as in case of secondary and tertiary sector the per capita income increased by 131 percent and 113 percent respectively.

Table 3
Per Capita NSDP in Different Sectors in Gujarat– (Rs. at 1980-81 prices)

Year	*Primary sector*	*Secondary sector*	*Tertiary sector*	*All sectors Gujarat*	*Per capita GDP in India*
1980-81	792	528	620	1,940	1,625
1990-91	729	914	997	2,641	2,267
1991-92	591	792	998	2,381	2,226
1992-93	867	1,132	1,093	3,091	2,298
1993-94	646	1,123	1,174	2,944	2,391
1994-95	896	1,379	1,255	3,531	2,518
1995-96	681	1,488	1,347	3,517	2,648
1996-97	779	1,577	1,443	3,799	3,915
1997-98	772	1,667	1,537	3,976	4,101
1998-99	741	1,047	1,171	3,043	4,371
1999-00	546	1,131	1,258	3,026	4,649
2000-01	491	1,222	1,320	3,094	4,928

Source: Directorate of Economics and Statistics (1997-98, 2001-02). From Table 6 (Hirway and Mahadevia, 2004: 25).

On the whole, Gujarat's economy has undergone significant structural transformation with the share of the non-primary sectors having significantly increased. Also, agriculture in the state is highly unstable, which tends to leave the small farmers and farm labour in poverty. (Hirway and Mahadevia, 2004: 25).

Employment and Labour Productivity in Gujarat

Table 4 below districts that, there is relative decline of primary sector both in terms of share in the Sate GDP as well as employment while share of secondary and tertiary sectors in GDP as well as employment are increasing.

The decline in the sectoral growth rate of employment and income in primary sector was due to the increasing pressure of population on land and instability in agriculture. This further led to casualisation of labour force in the rural areas. The micro studies and the Census data suggest substantial occupations diversification away from agriculture in rural Gujarat (Basant, 1993). The Census data also reflects that in the past few decades there has been a shift towards manufacturing, construction, trade and services.

The increase in the share of employment in the secondary sector in 1987-88 was due to increase in employment in the construction sector, which mainly constituted persons, engaged in the scarcity relief programmes set up by the government in the drought year 1987-88. The limited scope of generation of employment in the secondary sector is due to emergence of capital-intensive chemical based industry and decline in labour intensive industries like the textile industries.

The marginal increase in the growth rate of employment in service sector demonstrates the growing role of this sector and transfer of labour from agriculture to service sector.

Table 4
Income, Employment and Labour Productivity in Gujarat

Income	*% Shares in GDP* 1983-84	1987-88	1993-94	*Growth rate %* 1983-84 to 1987-88	1987-88 to 1993-94	1983-84 to 1993-94
Primary	36.8	18.3	21.6	- 18.6	10.5	- 1.1
Secondary	31.7	39.9	39.9	4.6	7.7	6.5
Tertiary	31.5	41.8	38.5	6	6.3	6.2
All Gujarat	100	100	100	- 1.1	7.7	4.2
Employment	*1983*	*1987-88*	*1993-94*	*1983 to 1987-88*	*1987-88 to 1993-94*	*1983 to 1993-94*
Primary	65.7	56.1	59.3	- 0.8	1.5	0.6
Secondary	15.1	23.9	19.2	5.9	- 0.4	2.1
Tertiary	19.2	20	21.5	1.5	1.6	1.6
All Gujarat	100	100	100	0.9	1.1	1.1
Labour Productivity	*1983*	*1987-88*	*1993-94*	*1983 to 1987-88*	*1987-88 to 1993-94*	*1983 to 1993-94*
Primary	3568	1816	2761	- 16.9	6.9	- 2.6
Secondary	13426	9294	15699	- 9.2	8.7	1.6
Tertiary	10444	11629	13581	2.7	2.6	2.6
All Gujarat	6373	5570	7573	- 3.4	5.1	1.7

Source: GOG (1998), NSSO, Sarvekshana (1981, 1986, 1988, 1990a, 1990b).

The macro trend shows that there is a shift of workforce from rural areas to urban areas, with one out of every three agriculture workers now in the manufacturing sector (Alagh1995: 118).

The workforce is getting absorbed mainly as casual labours and self-employed workers in urban areas. Also there is diversification of labour into construction and trade. However the incidence of poverty among casual and self-employed workers is higher than among the regular workers (Rani, 2004).

Work Participation in Gujarat

According to the Census 2001, there are 212.47 lakh work forces in the state in a population of 506.56 lakh. Of these, 170 lakh are main workers (79.8 per cent) and 43 lakh (20.2 per cent) are marginal workers (Table 5).

Table 5
Workforce in Gujarat

	1981		*1991*		*2001*	
	Population (in lakh)	*% of total population*	*Population (in lakh)*	*% of total population*	*Population (in lakh)*	*% to total population*
Population	341.00	-	413.09	-	506.56	-
Total Workforce	127.00	37.24	166.20	40.23	212.47	41.94
Main Workers	109.86	32.22	140.95	34.12	170.22	33.60
Marginal Workers	16.90	4.96	25.25	6.11	42.25	8.34
Non Workers	214.00	62.76	246.89	59.77	294.09	58.06

Source: Population Censes. [From Table 12 (Hirway and Mahadevia, 2004: 33)].

A comparison of the 1981, 1991, and the 2001 censuses show that the workforce participation rate (WPR) in the state has consistently increased from 37.24 per cent in 1981 to 40.23 per cent in 1991 to 41.94 per cent in 2001. The share of marginal workers in total workforce has also increased from 13.33 per cent in 1981 to 15.20 per cent in 1991 to 19.96 per cent in 2001, with the result that the WPR of main workers has declined marginally from 34.1 per cent in 1991 to 33.60 per cent in 2001. (Hirway and Mahadevia, 2004: 33).

The female work participation rate (WPR) in Gujarat is 39.0 per cent (2001) or 41.3 per cent (1999-00) in rural areas which is much higher than that in rural India (29.9 per cent in 2001). The female WPR in urban Gujarat, however, is marginally lower (13.5 per cent in 1999-00) than the all India rate of 13.9 per cent (Hirway and Mahadevia, 2004: 35).

Distribution of Main Workers in Gujarat

Gujarat is in a relatively better position than the country with respect to overall employment. As regards diversification of the workforce in the state; the share of agricultural employment (cultivators + agricultural labourers) has declined rapidly from 68.64 per cent in 1981 to 56.30 in 1991 to 52.05 per cent in 2001 (Table 6)–a decline of 16.59 percentage points in two decades.

Table 6
Distribution of Main Workers by Industrial Categories in Gujarat

Category	*1981*	*1991*	*2001*
Cultivators	41.45 (37.76)	47.04 (33.38)	46.93 (27.56)
Agricultural Labourers	25.13 (22.88)	32.31 (22.92)	41.71 (24.49)
Household industries	2.96 (2.71)	1.97 (1.40)	3.18 (1.87)
Other Workers	40.30 (36.71)	59.63 (42.30)	78.48 (46.08)
Total	109.84 (100.00)	140.95 (100.00)	170.30 100.00)

Note: Figures in lakhs. Figures in brackets indicate percentage of main workers.

Source: Population Censes. [From Table 14 (Hirway and Mahadevia, 2004: 34)]

The share of agricultural labourers in the agricultural workforce has increased from 37.68 per cent in 1981 to 40.71 per cent in 1991 to 46.97 per cent in 2001 (Hirway and Mahadevia, 2004: 34). On the whole, employment growth in the 1990s has been greater than the growth of the labour force (Table 7). During the year 1983 to 2000, the total employment and labour force increased by 36 percent, where as employment in private sector increased by 20 percent, employment in Public sector increased by 14 percent and employment in organized sector increased by 17 percent (Hirway and Mahadevia, 2004: 34).

Table 7
Labour Forces and Employment by Sector, Gujarat (1983-2000)

Employment (in lakh) Simple average annual growth rate (%)	*1983*	*1987-88*	*1993-94*	*1999-00*	*1983-94*	*1994-00*
Total labour force	133.18	149.05	167.83	207.36	2.6	3.9
Total employment	130.72	146.31	164.82	205.20	2.6	4.1
Public sector	8.00	9.15	9.75	9.32	2.2	-0.7
Private sector	6.39	6.47	7.20	7.95	1.3	1.7
Organized sector	14.39	5.62	16.95	17.27	1.8	0.3

Source: (i) The estimates of labour force and total employment are obtained by applying the corresponding UPSS (Usual status principal + subsidiary status) workers' proportion of various NSS rounds to the projected figures of respective years.

(ii) Employment figures of public and private sectors are from the Directorate of Employment and Training. From Table 15 (Hirway and Mahadevia, 2004: 34).

Employment Status of Workforce

The share of self-employed is about 59 per cent in rural areas and 41 per cent in urban areas (1999-00). The share of casual workers has increased from 18.3 per cent to 24.9 per cent in urban areas and has declined only marginally, from 34.8 per cent in 1983 to 34.5 per cent in 1999-00 in rural areas with the share fluctuating during the period. Compared to other states, the share of casual labour is much higher in Gujarat (Table 8) (Hirway and Mahadevia, 2004: 34).

Table 8
Distribution of 1000 Employed Persons, Gujarat (1983-2000)

	Rural				*Urban*			
	1983	*1987-1988*	*1993-1994*	*1999-2000*	*1983*	*1987-1988*	*1993-1994*	*1999-2000*
Self employed	597	478	502	592	418	388	383	410
Reg.sal. emp.	55	88	68	63	399	421	409	341
Casual labour	348	434	430	345	183	191	208	249
Total employed	1000	1000	1000	1000	1000	1000	1000	1000

Source: NSS 38th, 43rd, and 55th Round UPSS estimates.
From Table 16 (Hirway and Mahadevia, 2004: 35)

Only 6.3 per cent of workers are regularly employed in rural areas, while the corresponding figure in urban areas is 34.1 per cent. This share declined significantly during the 1990s in both rural and urban areas. While overall employment in the state has improved, the share of regular employment has declined, particularly in the 1990s, and the share of casual employment has increased. The share of workers covered under social security schemes of the organized sector is much less than 10 per cent. (Hirway and Mahadevia, 2004: 35).

Marketable Skills of Workforce in Gujarat

Unemployment rate in Gujarat is lower than that in the country. Unemployment was 4.63 per cent (Current Daily Status), in 1999-00 compared to 7.29 per cent for the country. Gujarat ranks tenth among major states of India with respect to unemployment rate (Hirway and Mahadevia, 2004: 35).

The share of skilled workforce (with some skills – not necessarily acquired formally) is less than 10 per cent in rural areas and less than 15 per cent in urban areas (Table 9).

Table 9
Percentage Distribution of Workforce by Possession of Marketable Skills in Gujarat 1993-94

	Rural		*Urban*	
Possessing	*Male*	*Female*	*Male*	*Female*
No Skill	89.9	93.7	80.4	88.8
Some Skill	10.1	6.3	19.6	11.2
Total	100.0	100.0	100.0	100.0
Persons Sampled	183,464	172,835	109,067	99,283

Source: Department of Industries, Gandhinagar. [From Table 22 (Hirway and Mahadevia, 2004].

Educated Unemployment

The present situation, however, is that of a mismatch between the demand and supply of educated labour power in the state, with the army of educated unemployment increasing continuously along with serious shortages of technical manpower in several fields. The strength of the educated unemployed has more than doubled, and was estimated at 6.82 lakhs in 1998. Also, the share of the educated in the total unemployed has increased from 57 per cent in 1980 to 75 per cent in 1994, indicating the low and declining employment opportunity for the educated in the state (Hirway and Mahadevia 1999).

INCIDENCE OF POVERTY IN GUJARAT

As regards income poverty, Gujarat is above the all-India average. The incidence of poverty in Gujarat is 23.92 per cent (1993-94) as against 36 per cent in India. The incidence of poverty in Gujarat has declined considerably, from 46.35 per cent in 1972-73 to 22.18 per cent in 1993-94 in rural areas and from 41.21 per cent in 1972-73 to 27.07 per cent in 1993-94 in urban areas, which implies that economic growth has helped in poverty reduction (Table 10).

Table 10
Rural and Urban Poverty in Gujarat

	Number in Lakhs			*% to Total Population*		
Year	*Rural*	*Urban*	*Total*	*Rural*	*Urban*	*Total*
1972-73	94.61	41.09	135.70	46.35 (5)	41.21 (7)	47.21 (5)
1977-78	92.53	41.33	133.86	41.76 (5)	43.13 (6)	42.17 (6)
1983	73.49	47.26	120.75	29.8 (5)	40.63 (9)	33.27 (5)
1987-88	75.95	52.63	128.58	28.67 (5)	39.63 (8)	32.33 (6)
1993-94	62.16	41.77	103.93	22.18 (3)	27.07 (6)	23.92 (3)

Note: Figures in brackets indicate the rank of the state economy among the major 16 states in India.

Source: Planning Commission (1997) Estimates of Poverty in India, New Delhi.

Incidence of Poverty Across Regions in Gujarat

A region-wise view of poverty shows that the environmentally degraded regions i.e. the eastern tribal belt and the northern plains have the highest incidence of poverty. The eastern tribal belt is the poorest region in the state with regard to both rural and urban poverty (Table 11). The environmentally degraded northern plains are second from the top with respect to income poverty, incidence of poverty being 24.58 per cent in 1993-94.

Urban poverty is a relatively serious problem in the state, with the incidence of urban poverty being higher than that of rural poverty. The incidence of urban poverty declined only marginally during the 1980s, with the absolute number of the urban poor increasing from 41.33

lakhs in 1977-78 to 47.26 lakhs in 1983, and to 52.63 lakhs in 1987-88. The probable reasons cited for this were (a) Decline in the organized textile industry (the economy of some small towns was centred around a composite textile mill or two) leading to direct and indirect unemployment (b) Declining quality of employment arising from informalisation of the labour market (c) Urbanisation as a result of migration from rural areas. There has been a significant decline in the incidence of urban poverty in Gujarat, thanks to the increasing industrial investment. The incidence of poverty, however, is higher in smaller

Table 11
NSS Region-Wise Incidence of Poverty

	*Incidence of Poverty (%)**					
NSS Region	*1983 %*	*1987-88%*	*% Reduction 1983 to 1987-88*	*1993-94%*	*% Reduction 1987 to 1993-94*	*% Reduction 1983 to 1993-94*
			Rural Gujarat			
Eastern	61.40	34.19	-44.32	24.12	-29.45	-60.72
NorthernPlains	35.70	25.87	-27.54	20.52	-20.68	-42.53
SouthernPlains	36.60	22.85	-37.57	23.51	2.89	-35.77
Dry Regions	46.80	46.95	0.32	22.52	-52.03	-51.89
Saurashtra	23.50	18.95	-19.36	10.03	-47.07	-57.45
			Urban Gujarat			
Eastern	39.30	39.32	0.05	34.33	-12.69	-12.65
NorthernPlains	31.30	34.23	9.36	30.05	-12.21	-4
SouthernPlains	41.70	30.89	-25.92	20.89	-32.37	-50
Dry Regions	51.10	53.88	5.44	27.03		
Saurashtra	41.50	53.77	29.57	34.04	-36.69	-17.98
			Total			
Eastern	na	34.49	na	25.06	-27.34	na
Northern Plain	na	29.03	na	24.58	-15.33	na
SouthernPlains	na	25.85	na	22.45	-13.15	na
Dry Regions	na	40.20	na	23.30		na
Saurashtra	na	28.18	na	18.80	-33.29	na

Source: DES (2000), Government of Gujarat, Gandhinagar.

towns, largely because of the low employment avenues, and the predominance of low skills/productivity of work, low wages and poor infrastructural facilities in smaller towns.

In terms of absolute numbers, there is decline in poverty by about 23 per cent in the last two decades i.e. from 13 million in 1973-74 to 10 million in 1993-94 (Hirway 2000: 3116). There is a substantial decline in the incidence of poverty by 32 percent in urban areas during 1987-93 against 23 percent decline in the incidence of poverty in rural areas. This indicates that the resilience of the urban economy due to high growth of investment and employment opportunities in the secondary and tertiary sectors (Hirway 2000: 3118).

More Poverty in Urban Areas

The urban poverty was higher than the rural poverty in all regions of the state except in the southern plains (Table 12). This decline in rural poverty could be due to sectoral diversification in rural areas of the state, which is reflected in non-household based manufacturing and other non-farm activities (Basant *et al.*, 1998).

Table 12
Incidence of Rural and Urban Poverty in Gujarat[2]

(In percentage)

Regions	*1987-88*			*1993-94*		
	Total	*Rural*	*Urban*	*Total*	*Rural*	*Urban*
Gujarat Eastern	34.49	34.19	39.32	25.06	24.12	34.33
Northern Plains	29.03	25.87	34.23	24.58	20.52	30.05
Southern Plains	25.85	22.85	30.89	22.45	23.51	20.89
Gujarat Dry Regions	40.2	46.95	53.88	23.3	22.52	27.03
Saurashtra	28.18	18.95	53.77	18.8	10.03	34.04

Gujarat is one among the worst affected states during droughts and has been receiving funds under the drought relief programmes. Also a large segment of the poor migrating from the rural areas to the urban areas are getting absorbed in low productivity tertiary activities where employment has increase in the recent years (Kundu 2000).

Poverty in Rural Areas

The primary sector, which consists mainly of agriculture, is highly unpredictable due to vagaries of monsoon. The manufacturing sector in Gujarat developed due to this precarious state of agriculture with the justification that manufacturing is the growth engine in arid and semi arid regions (Rani, 2004).

Agriculture in Gujarat is primarily rainfed. However the State has recorded progressive improvement in the overall levels of agricultural production, which was possible on account of the increase in productivity in some pockets of Gujarat. This was also possible due to increased use of chemical fertilizers and the extension of areas under high yielding varieties in irrigated areas of the state.

According EPW, January 13, 2002 in Gujarat 624170 pumpsets (88.21 %) were energized by March 1998. At all India level 81.72% of 119.46 lakh pumpsets were energized. With about three-fourths of the cultivated area boing dependent on the vagaries of the monsoons, the Green Revolution, which needs at least some assured water supply, has a limited coverage in the state.

The agricultural population from dry regions, particularly the northern dry plains and the eastern tribal region, migrates after the rainy season to distant places in search of employment. Agricultural population, particularly the labourers, suffers from low and fluctuating incomes on the one hand, and poor access to social and economic programmes due to migration on the other.

AGRICULTURE IN GUJARAT

65.30 % of the total area of Southern Gujarat Agro Climatic region is irrigated through canal which is the maximum among all the agro climatic regions. The minimum area irrigated through canal is the South Saurashtra Agro Climatic Region (Table 13).

Table 13
Irrigation in Gujarat (1994-95)

Regions	*Area (000 Ha.)*	*Irrigation (%)* Canals	Tanks	Ground Water	*Groundwater Utilization (%)*
Southern Hills	742	46.11	1.63	52.26	21.02
Southern Gujarat	1557	65.30	0.51	34.19	20.79
Middle Gujarat	2351	29.28	3.73	66.99	30.01
North Gujarat	3782	11.94	0.31	87.75	56.16
North-West Arid	4565	14.27	0	85.73	30.28
North Saurashtra	4824	10.75	1.96	87.29	33.86
South Saurashtra	1056	6.22	0	93.78	51.27
State	18877	20.97	1.36	77.67	37.40

Agricultural growth in Gujarat has been largely supported by expanding irrigation facilities that depend predominantly on groundwater. The misappropriate use of groundwater, without any proper recharge, has led to the over-exploitation of this resource. The irrigation policy and subsidies, the availability of cheap diesel and electricity, accompanied by a strong rich farmers' lobby, have all aided this over-exploitation.

Even today, the process continues, making for non-sustainability of agricultural growth in the state. In short, agricultural growth in the state in the recent years has been low and unsustainable, with a narrow base which does not seem to be conducive to rapid poverty reduction for the agricultural population.

Distribution of Land Holdings

There has been an increase in the number of land holding in case of Marginal, Small and Semi–medium groups as well as area operated under these land holdings groups in Gujarat as well as in all India (Table 14).

However the increase is more in Gujarat when compared to all India. There has been a decline in the number of land holding in case of medium and large groups as well as area operated under these land holdings groups in Gujarat as well as in all India. However the decrease in number of large land holding is more in Gujarat when compared to all India.

Table 14

Distribution of Land Holdings[3]

* No. of holdings & Area operated in '000 Nos

Group	*Particulars**	*India* 1985-86	1990-91	% Change	*Gujarat* 1985-86	1990-91	% Change
Marginal	No. of Holdings	56147	62106	11	801	924	15
	Area Operated	22042	24616	12	416	489	18
Small	No. of Holdings	17922	19962	11	737	915	24
	Area Operated	25708	28701	12	1080	1343	24
Semi- me-	No. of Holdings	13252	13913	5	785	890	13
dium	Area Operated	36666	38350	5	2233	2515	13
Medium	No. of Holdings	7916	7630	-4	678	669	-1
	Area Operated	47144	45052	-4	4126	4005	-3
Large	No. of Holdings	1018	1668	-13	145	118	-19
	Area Operated	33002	28899	-12	2099	1941	-8
All Sizes	No. of Holdings	97155	105278	8	3145	3517	12
	Area Operated	164562	165619	1	9954	10292	3

INDUSTRIES IN GUJARAT

The State could attract industries and it ranks second in terms of investments in projects. The State has also taken various initiatives to invite private sector participation in the infra-structural projects like power, ports, roads, etc. and the industrial structure has been gradually diversifying with the development of industries like chemicals, petrochemicals, fertilizers, engineering, electronics, etc. The percentage share of Gujarat State in All India aggregates in the year 1997-98 is mentioned hereunder:

Number of factories	:	9.8 per cent
Number of employees	:	8.71 per cent
Productive capital	:	15.27 per cent
Value of output	:	15.27 per cent
Net value added	:	11.45 per cent

Gujarat State ranked second in respect of state-wise percentage of share of both, in Gross Value of Output as well as in Net Value Added by manufacture generated by factory sector of the country.

The following are the sectors, which registered growth in 1998-99 over 1997-98 at constant (1993-94) prices:

Mining and quarrying	:	13.08 per cent
Registered manufacturing	:	5.71 per cent
Electricity, Gas and Water supply	:	6.36 per cent
Transport	:	5.77 per cent
Communication	:	18.64 per cent

The number of registered working factories in the State has increased from 19574 in 1998 to 19771 in 1999. The average daily employment in these working factories, which was 8.60 lakh in 1998 increased to 8.65 lakh in 1999.

Gujarat has witnessed impressive growth in the small-scale sector. There was 2160 small-scale industrial unit in 1961. It has increased to 15849 in 1970, 43712 in 1980 and 1,15,384 in 1990 and 2,33,777 by October, 1999.

Growth in Manufacturing Sector

Gujarat ranked eighth in industrial development among the major states in India in 1960, the year when the state was formed. Today the state is almost at the top with regard to industrial development. The industrial sector has performed consistently well, with the growth rate of the manufacturing sector continuously increasing from 3.04 in the 1960s to 5.55 in the 1970s, 7.73 per cent in the 1980s and above 9.00 in the 1990s. The major factors which helped the growth of industries here are the establishment of industrial estates with all the required infrastructural facilities in several parts of the state by the GIDC (Gujarat Industrial Development Corporation), the continuous availability of power and water to industry, the availability of railway links with major industrial and trading centres (particularly in south Gujarat where the industries are concentrated), the setting up of four important corporations, namely, GIIC (Gujarat Industrial Investment Corporation), GIDC, GSFC, (Gujarat State Finance Corporation) and Indextb (Industrial Extension Bureau) in order to provide finance, power facilities and all the required support to industries, and the highly positive and supportive attitude of the government to incoming industries. The industrial base of Gujarat has gradually and increasingly diversified with the establishment of petrochemicals, fertilizers, dyestuff, pharmaceuticals and drugs, as well as engineering and electronics industries in the state.

The process of industrialization in Gujarat has taken a new turn since the introduction of the economic reforms in 1991. After the Government of India announced the New Industrial Policy (NIP) in 1991 with the objective of implementing the economic reforms in the industrial sector, the Government of Gujarat also responded favourably and announced its own industrial policies. The State Industrial Policy 1990-95 was already in operation when the NIP 1991 was declared by the Central Government. The Government of Gujarat thereafter declared the 'New Industrial and

Incentive Policy 1991-2000' and 'Gujarat: 2000 AD and Beyond' in 1994 for accelerated industrial development of the state.

Although the policy declarations by the Government of Gujarat are essentially based on the general framework of the NIP 1991, they are more aggressive in terms of promoting and facilitating new industrial investments. According to the Government of Gujarat, the new climate of economic liberalization and globalisation has opened up new opportunities to the states in India to attract industrial investments from India and abroad. The state's approach is 'to compete not only with other Indian states, but also with the newly emerging high growth regions of South-east and East Asian countries in the industrial sector' (Indextb 1996).

As in the case of the Central Government, the main policy instruments of the state government are deregulation and liberalization of the different markets, incentives and concessions to potential industrialists from India and abroad, and promotional and developmental work. The focus of the policy, however, has been relatively more on incentives and concessions, and on promotional and developmental work. There is also a clear emphasis on promoting larger units which are described as 'premier' (units with investment between Rs.100 crore to Rs.500 crore) and 'prestigious units' (units with investment of more than Rs.500 crore) in the policy statements.

The state government has set up the Gujarat Infrastructure Development Board to promote and support the development of critical infrastructural facilities in the state; to give special concessions and subsidies as well as help in the acquisition of land and water to incoming industries; and to offer special benefits to export-oriented industries as well as to Non Resident Indians. The government has also set up a special cell in the Indextb for providing all required information and assistance to potential industrialists.

Increasing Flow of Investment

This approach has paid rich dividends in terms of attracting industrial investments to the state, particularly in the large and medium industries. In fact, Gujarat has attracted the highest industrial investment in the large and medium sectors, which was more than Rs.1,70,000 crore during 1991-97 among all the states in India. This quantum jump implies an increase of more than 10 times in the industrial investment per year, a five-and-a-half times increase in the number of projects per year, and an increase of more than 60 per cent in the average investment per unit, from Rs. 24 crore to Rs. 40 crore.

The new industrial investments have also gradually introduced the following set of new characteristics in the industrial scene in Gujarat: (a) There is a clear increase in the investments in large and medium industries in the state, with the small sector lagging behind in the post-1991 period, (b) The location pattern of industries has, to an extent, shifted to coastal Saurashtra, with 21 per cent (instead of 2 per cent before 1991) of the new investments located in the region; (c) Although the basic pattern of industrial dispersion has not changed, as the industrially rich regions have attracted larger industrial investments, several small centres have also developed/strengthened in the post-reform period; (d) The new investments are much more capital-intensive than before, particularly in the Saurashtra and Kachchh regions; (e) The share of mineral-based industries is very large (9 per cent); (f) There is an export orientation in the new investments; and (g) The share of pollution-prone industries and industries using non-renewable resources has remained high (Hirway and Shah 2000).

This quantum jump in industries, and particularly medium and large-scale industries, is likely to benefit the hinterland only if the land and labour markets function well. The labour market can benefit the hinterland if industries generate enough employment directly and indirectly, the local labour is skilled and professionally trained to meet the demand generated by these industries, and labour laws ensure a reasonable quality of employment in terms of wages, social protection and health protection.

Negative Consequences for Labour and Livelihoods

The new pattern of industrialization does not seem to be very positive in these respects: (a) These industries are highly capital-intensive, and are not likely to generate enough employment for the unemployed (Patel 1999); (b) The local labour may not get quality employment in these industries due to the low level of literacy, particularly in rural areas. Since the efforts undertaken to train the local population to use them in industries are not adequate, most professional and skilled workers for industries are likely to come from urban centres in and outside Gujarat. As a result, the local labour is likely to get only casual/contractual unskilled work which tends to be irregular, low-paid and almost without any social protection (Hirway and Shah 2000).

In the case of the land market as well, the policy of the state government that offers land to industries on a platter on the one hand, and the distorted nature of the land market (due to the rules and regulations in the land market) on the other, are not likely to allow the sellers of land to get the right prices. Our study in Saurashtra has shown that (a) small and marginal farmers, particularly those belonging to low castes, usually sell their entire land to industries; (b) the farmers get low prices for the land and (c) the compensation/remuneration does not ratify any livelihood for the future. In short, the acquisition of land by industries tends to affect the livelihood of the poor adversely.

Another important issue concerns the model of industrial development that the state seems to have adopted. The question is whether it is sustainable and whether it can generate enough employment opportunities in the economy. Theoretically, there are two approaches to industrialization, namely, the Textile First approach and the Machinery First approach. While the former approach advocates highest priority to industries that produce non-durable consumer goods like textiles, sugar, tea, edible oil, light engineering goods, intermediate goods and simple equipment and machines, the latter advocates heavy capital goods, basic inputs etc. The former tends to strengthen the industry-agriculture linkages, and generates large-scale employment as well as higher growth rates in the short run. The question therefore is whether there is a need to push the second strategy through heavy incentives and concessions, and promote these industries even if they do not seem to benefit the state in the near future.

There is an ongoing race among some major states in the country in providing incentives to attract industrial investments to their respective states. This race is not only market friendly but also anti-market in character, and therefore likely to give the wrong signals if not managed carefully; it reduces the revenue of the state drastically, which in turn reduces allocations to the social sectors, and such incentives to new units are unfair to the already existing units. Recently, some steps have been taken to make these rates uniform for all the states. However, it remains to be seen how effectively these are formalized and implemented in an environment of tough competition for attracting industrial investment.

ENVIRONMENTAL CONSEQUENCES

Pollution and degradation of natural resources, degradation of forests and relative neglect of tribal areas are considered.

Pollution of Land, Air and Water Resources

Pollution-prone industries are predominant in Gujarat partly because of some historical reasons, such as the growth of textile-related chemical industries, discovery of oil and natural gas, and the growth of drug and pharmaceuticals industries in the state in the pre and post-independence period. Recently, there has been a quantum jump in mineral- based industries as well. However, the pollution generated by these industries has not been controlled adequately, with the result that the workers of these units as well as the people living in surrounding areas suffer from serious health hazards. Although the state government has initiated efforts in promoting knowledge-based industries recently, not much has been achieved so far. It is also relevant to see who pays the cost of the pollution. Studies have shown that the victims of the adverse impact of pollution are industrial workers, particularly those working in small-scale units as casual and contract workers, and the vulnerable population in the surrounding areas.

Water Availability and Unsustainable Usage

There is a general belief that the water shortage in Gujarat is caused by nature. However, this is not quite true. In spite of the low and erratic rainfall, the combination of climate, physiography and geology in different regions of the state did provide somewhat favourable conditions for water resources in most regions of the state about three decades ago (Hirway and Patel 1994). For example, the north Gujarat alluvial area has low rainfall, but also has good topographic conditions of recharge and ideal conditions of aquifers which have rendered the region rich in groundwater reservoirs. In fact, this storage has supported thousands of tube wells for more than 35 years in this region! Also, although the hilly areas of eastern Gujarat have adverse physiographic and geological conditions that largely inhibit groundwater storage, they also have ideal sites for creating surface storage dam reservoirs. The arid areas of Kachchh have favourable geological formations, which have provided confined aquifers in consolidated formations of sweet (ground) water upto 200 m depth. Finally, the coastal areas of Saurashtra are capable of storing the rainfall run-off from the upland rocky terrain, thereby naturally conserving the precipitation of the region.

The shortage of water experienced by the state is largely man-made, and caused by mismanagement of water resources. Coastal Saurashtra enjoyed a sweet (ground) water regime, but the overdrawing of groundwater has pushed in saline water, resulting in salinity ingress in the region on the one hand, and deterioration in the quality of water supply on the other. The indiscriminate withdrawal of groundwater in north Gujarat, Kachchh and northern Saurashtra has depleted water resources badly, leading to water mining on a large scale. Even in south Gujarat, where the rainfall is high, canal irrigation and the consequent overuse of water have increased waterlogging and salinity in this region. This mismanagement, primarily arising from the pricing and subsidy policy of irrigation, has resulted in water becoming one of the most scarce items in the state.

The mismanagement of water resources has also led to deterioration in the quality of water available in different regions of the state. The most important quality problems pertain to excessive salinity, excessive fluoride and excessive nitrites, all of which have resulted in a severe shortage of drinking water in Gujarat. About half the villages in the state face a serious shortage of drinking water in summer. Although transfer of water supply from one place to another through regional pipeline schemes is being attempted in some regions, it has not worked well. The shortage of drinking water has serious implications for the well-being of the poor. Women travel long distances to collect water, at times spending more than four hours a day. The non-potable water results in several water-borne diseases such as fluorosis, diarrhoea, gastroenteritis etc. in the state, and the poor are the worst sufferers (Hirway and Mahadevia 1999).

Degradation of Forests and Neglect of Tribal Areas

The welfare of tribals (who form about 15 per cent of the state population and who are the poorest social group in Gujarat) is closely connected with the status of forests. The forest area in Gujarat covers about 10 per cent of its geographic area. The actual forests; however, cover only 6 per cent of the area (Forest Survey of India 1997). Although Gujarat experienced a loss of forest cover till about the 1970s, the decline has been arrested to an extent. Actual forest cover is stabilized around 63 m ha. In terms of the quality of forest, however, the state shows a declining trend, as the percentage of dense forests has consistently depreciated during the past two decades or so.

The improvement in mangroves in the recent years, after a consistent loss for decades, is limited only to a few regions around the Gulf of Kachchh. The degradation of forests as well as the ownership pattern of forests in the state has adversely affected the livelihood of tribals.

Degradation of land is another major problem in Gujarat. Deforestation and degradation of forests in the eastern belt, salinity ingress in the coastal region, desertification in northern region and general soil erosion have increased the area under degraded land in the state (Gujarat Ecology Commission 1997).

The fact that about 54 per cent of the talukas in the state are under the special areas development programmes like Drought Prone Area Programme or Desert Development Programmes indicates the severity of land degradation in the state. Once again, the poor are the worst sufferers, as these degraded lands usually belong to small and marginal farmers or to common lands which provide significant support to the livelihood of the poor (Hirway and Mahadevia 1999).

Land degradation implies low land productivity; low employment avenues and seasonal migration of the poor in search of work. All of these tend to enhance their poverty.

DEVELOPMENT PATTERNS IN GUJARAT

Two parameters which matter ultimately to any society, namely, life expectancy at birth and infant mortality rate along with the state expenditure on health were examined.

Life Expectancy at Birth (LEB)

Life expectancy at birth and infant mortality rate are known to be two robust indicators of the health of a population. Gujarat ranks low on both these indicators. In the case of LEB, the rank of the state is ninth (62.9 years as against 62.7 in India) in 1991and the rank is sixth (63 years against 73 in India) in 1994-96. In the case of male LEB the state ranks eighth (61.7), while in the case of female LEB, the state stands ninth (64.3). LEB in rural areas is much lower (62.1) than the same in urban area (64.6), which implies that living conditions and living standards in rural areas are worse than urban areas.

Gujarat has recorded an increase of 17.8 per cent in its LEB during 1971-91, which is more than the increase experienced at the all-India level (11.07 per cent). The rate of increase in LEB, however, has declined during 1981-91, from 11.80 per cent during 1971-81 to 5.36 per cent in 1981-91. The decline is observed in rural and urban areas, and both for men and women (Table 15).

Table 15
Life Expectancy at Birth in Gujarat

		% of increase			
Particulars	*1971*	*1981*	*1991*	*1971-81*	*1981-91*
Gujarat Total	53.4	59.7	62.9	11.80	5.36
Male	54.8	59.2	61.7	8.03	4.22
Female	52.0	62.3	64.3	19.81	3.21
Rural LEB	52.4	59.0	62.1	12.60	5.25
Urban LEB	52.7	62.0	64.6	17.65	4.19

Source: Registrar General of India (1996), Vital Statistics of India 1990. Based on The Civil Registration System, Ministry of Home Affairs, New Delhi.

Infant Mortality Rate (IMR)

The IMR in Gujarat is 63 (1994-96) as against 15 in Kerala! Also, the IMR in rural Gujarat (69) is more than 40 per cent higher than that in urban Gujarat (48). The IMR in Gujarat declined by 55 per cent during 1971-96 (56.05 per cent in rural areas and 54.72 per cent in urban areas), with the state ranking second in the rate of decline among the major states of India. The first position was occupied by Kerala, whose IMR had reduced by 75 per cent during this period.

The rate of decline in the IMR in Gujarat seems to have decelerated in the recent years. This phenomenon has been observed for many other states in India also (Hirway and Mahadevia 1999). Table 16 shows that the IMR has remained stagnant in the state in 1990s, with the urban IMR declining only marginally and the rural IMR remaining same.

Table 16
IMR in Gujarat
(3-yearly moving averages)

	1971-73	*1981-83*	*1991-93*	*1996-98*
Total IMR	144	111	64	64
% change	–	-22.92	-42.34	0.0
Rural IMR	157	121	70	69
% change	–	-22.93	-42.15	-1.43
Urban IMR	106	83	51	48
% change	–	-21.70	-38.85	-5.88

Source: Registrar General of India (1996), Vital Statistics of India 1990. Based on The Civil Registration System, Ministry of Home Affairs, New Delhi.

In short, there has been a deceleration in the performance of the two health indicators in the recent past. Although no firm reasons have been presented by scholars, it has been observed that the fault does not lie only with direct efforts at the state level. The health status of a society depends on two sets of factors, namely, health and medical services of different kinds and the general nutrition level of the population. The latter is influenced by factors such as poverty and unemployment levels, environment and pollution status, women's position in the family; etc. which are macro variables influenced by the development path of the economy. For example, a decline in IMR takes place not only due to improved neo-natal facilities, better immunisation, extension of health care system etc., but also when tile health status of mothers improve due to better nutrition facilities provided to girls, marriage at an optimum age, optimum age at the time of first child-birth, spacing of children and the general improvement in the status of women, including their improved literacy. The initial gains in health occur due to direct efforts, but sustainable long-term gains are achieved through macro factors determined by the development path that the state has chosen.

Expenditure Pattern in Health Sector, Gujarat

In the case of the health sector the composition of expenditure has changed in such away that a relatively higher amount is spent on the medical side (Table 17). As against this, expenditure on public health as percentage of total health budget shows a noticeable decline while expenditure on family welfare has drastically decreased. The share of water and sanitation in the health sector has increased. (Hirway and Mahadevia, 2004: 65).

Table 17
Expenditure Pattern in Health Sector, Gujarat

	Percentage of expenditure on			
Year	*Medical*	*Public health*	*Water & Sanitation*	*Family welfare*
1985-86	41.40	16.30	24.80	17.60
1990-91	51.10	11.60	24.60	12.70
1995-96	53.65	15.19	17.86	13.29
1996-97	56.50	15.51	15.64	12.34
1997-98	57.22	14.10	22.06	6.62
1999-00	37.48	8.31	41.32	12.89
2000-01	30.39	5.70	57.04	6.87
2001-02	49.17	9.34	32.73	8.76

Source: Prabhu (1997), Directorate of Economics and Statistics (1999) and Budget Publications No. 1,11,18 and 24, for the years 2001-02, 2002-03 and 2003-04. From Table 4 (Hirway and Mahadevia, 2004: 65)

Notes

1. *Source*: Socio-economic Survey Volumes from 1985-86 to 1997-98, Directorate of Economics and Statistics, Government of Gujarat, Gandhinagar.
2. *Source*: Table 15 from Hirway (2000: 3116)
3. *Source*: Ministry of Agriculture, Agricultural Situation in India (August, 1995), Directorate of Economics and Statistics.

4

DISPARITY IN SELECTED RURAL URBAN CLUSTERS IN GUJARAT

RATIONALE FOR SELECTION

In accordance with the criteria decided by the study team for selection of urban areas and on the basis of information on population, percentage of people engaged in primary, industrial and services sector (Table 1), the Worker Population of the all the urban areas in Gujarat were examined. Keeping in view, the districts where urbanization index is less than 25 according to 1991 Census, the following three urban areas are selected for conducting the study. These three districts also belong to three different regions in Gujarat-South Gujarat, North Gujarat and Saurashtra regions– apart from different agro-climatic zones. The three different urban areas selected for the study also have different sectoral composition in terms of relative importance of primary, secondary and tertiary sectors of the economy. The criteria used are the percentage of work population engaged in primary, industrial and service sectors. The brief socio-economic profile of the districts and the respective towns is presented hereunder:

1. **Predominance of trade and services:** Rajpipla, Narmada District-South Gujarat region; Southern Gujarat agro-climatic region;

2. **Predominance of industrial activity:** Thangadh, Surendranagar District-Saurashtra region; North Saurashtra agro-climatic zone;
3. **Predominance of agricultural commodity trading:** Unjha, Mehsana District-North Gujarat region; North Gujarat agro-climatic region.

Table 1
Sector wise Worker Population in Selected Urban Areas

Major Towns	*Population '000*	*Male '000*	*Female '000*	*Density*	*Primary (%)*	*Industrial (%)*	*Services (%)*
Rajpipla	33.11	17.02	16.10	7441.12	21.0	16.0	63.0
Thangadh	24.77	12.95	11.82	491.57	8.0	61.0	31.0
Unjha	51.00	26.38	24.62	1451.00	NA	NA	NA

TRADE AND SERVICES IN NARMADA DISTRICT

The Narmada district is situated between the bank of the Narmada and the bank of the Karjan and is one of the tribal districts of South Gujarat Region. The total area of this district is 3,73,699 hectares, including the forest area of 1,17,377 square kms. The total population of the district is more than 5 Lakh, as per census 2001.

There are four Talukas in this district i.e. Nandod, Dediapada, Sagbara, and Tilakwada taluka. Only one town has the honour of having Municipality, all other villages have Gram Panchayats. There are almost 630 villages. Most of areas of the district are tribal areas. This is a fast developing district. There is a Gujarat Spinning limited near village Arnletha and the Narmada Sugar near village of Dharikheda. There is Prashant Pharmacy, which manufactures the Ayurvedic Medicines, and there is also one Polythene bag factory in this district.

Profile of Rajpipla–Trade & Services Cluster

As of 1991, Rajpipla town has 3763 enterprises and 16% of total labourers and 23% of labourers are engaged in agriculture and trade and commerce related activities respectively. This city also seems to be entertaining trading and commerce related activities. There is a possibility that the settlements in the vicinity are dependent on Rajpipla market for selling their produce and buying inputs and consumption goods. Also being a tribal dominated district, there is every possibility that due to lack of infrastructure, the market is concentrated in this town. Also looking at the commodities imported, manufactured and exported, the town seems to be catering to agriculture and trading related activities. In this backdrop we have selected Rajpipla town as one of the three towns of Gujarat State.

Imported: Cloth, food-grains and Edible oil
Manufactured: Carbon, Ayurvedic medicines and tiles
Exported: Bananas, sugarcane and tur dal

Profile of Selected Villages

For conducting the study, the following six villages[1] in the vicinity of Rajpipla Town has been selected:

ii.	Rampara (Mangrol)	ii.	Thari
iii.	Vavdi	iv.	Karantha
v.	Lachhras	vi.	Mangrol

DEMOGRAPHIC PROFILE

As per the 2001 Census, the number of households in the six selected villages varies from 170 to 1000 with Rampara (Mangrol) village having the minimum number of households i.e. 171 and Lachhras village having the maximum number of households i.e. 987. The population in the selected villages varies from 780 to 4500 with Thari village having a minimum population of

783 and Vadia village having the maximum population i.e. 4480 (Table 2).

Table 2
Demographic Profile of the Selected Villages and Rajpipla Town

(As per 2001 Census)

Particulars	*Raj-pipla (M)*	*Ram para (Man-grol)*	*Thari*	*Vadia (CT)-*	*Karan-tha*	*Lachh-ras*	*Man-grol*
Total Area of the village (Ha.)*	4.45 Sq.km.	315.81	270.34	591.42	764.9	1027.88	1127.53
No. of households	7034	171	159	888	273	987	285
Total population	34923	806	783	4480	1374	4006	1535
Avg. Size of HH	4.96	4.71	4.92	5.05	5.03	4.06	5.39
Sex Ratio	953	1025	997	710	927	917	744
Total population of children in Age group (0-6 Yrs.)	3699	92	82	324	164	508	172
Population of age group (0-6 Yrs.) as % of total population	10.59	11.41	10.47	7.23	11.94	12.68	11.21
Scheduled Caste	1638	0	17	361	110	165	127
SC (% of total population)	4.69	0.00	2.17	8.06	8.01	4.12	8.27
Scheduled Tribe	7560	621	313	2147	513	2394	604
ST (% of total population)	21.65	77.05	39.97	47.92	37.34	59.76	39.35
Literates	26673	418	499	3708	865	1823	972
Literates (% of	76.38	51.86	63.73	82.77	62.95	45.51	63.32

* As per 1991 Census.

The sex ratio in the selected villages varies from 710 to 1025. Among the six villages selected, the sex ratio is lowest in Karantha village and highest in Rampara (Mangrol) village. The population in the age group of 0 – 6 years varies from 7% (Vadia village) to 12.68% (Lachhras village) of the total population.

The percentage of total population of Schedules Caste is maximum in Mangrol village and is nil in Rampara (Mangrol) village. The population of Schedule Tribes is significant in the selected villages with maximum in Rampara (Mangrol) village (77% of the total population) and minimum in Karantha village (37% of the total population).

The literacy rates in the selected villages vary from 45.51% (Lachhras village) to 83% (Vadia village).

Amenities

There is Primary or Elementary education facility in all the selected villages, Matriculation and Secondary School in Lachhras, Mangrol and Vavdi village and Adult Literacy Class/Centre in Rampara (Mangrol) village (Table 3).

Table 3
Amenities Available in the Selected Villages

(As per 2001Census)

Particulars	*Lachh-ras*	*Thari*	*Man grol*	*Ram-para (Man grol)*	*Karan-tha*	*Vadia CT*
Educational	*P(3),H*	*P(2)*	*P(2),H*	*P(2),Ac*	*P(2)*	*P(3),H*
Medical	MH, PHC, D, FPC	CHW	D, RP, SMP, CHW	CHW	PHS, CHW	PHS, RP, CHW
Drinking water (Potable)	T, W	T, W	T, W, R	T, W, TW, R	T, W, TK	T, W
Post & Telegraph	PTO, Phone	PO, Phone	PO	–	–	PO, Phone
Day or Days of Market/Haat, if any.	–	–	–	–	–	–
Communication (Bus Stop, Railway Station, Water Way)	BS	BS	BS, NW	BS, NW	BS	BS
Approach to village	PR, KR	PR	PR, KR, NW	PR, KR, NR	PR, KR	PR
Nearest town (Distance)	Raj-pipla (16)	Raj-pipla (8)	Raj-pipla (15)	Raj-pipla (10)	Raj-pipla (6)	Raj-pipla (5)
Power supply	EA	EA	EA	EA	EA	EA

All the selected villages have a Community Health Worker/Health Worker or Dispensary or Primary Health Centre or Sub Centre or have Registered Private Practitioner. Maternity Home is present in Lachhras village.

There is potable drinking water facility in all the villages through tap water and well water. Out of the six villages selected, only four villages have either post office or post and telegraph office and phone connections available in only three of the six villages. All the villages have Pucca Roads and are connected with Bus Service. Mangrol and Rampara (Mangrol) has navigable water way. All the selected villages have electricity for all purpose.

WORKER POPULATION

The percentage of worker population (Table 4) to total population in the selected villages varies from 26 % (Vadia village which is relatively more prosperous) to 50 % (Lachhras Village which is remote and relatively less prosperous).

The population engaged as main workers in the selected villages varies from 22 % to 42 % with Karantha village having the maximum percentage of population (42 %) engaged as main workers and Vadia village has the minimum percentage of population (22 %) engaged as main workers. Among all the selected villages, Lachhras village has the maximum percentage of population engaged as marginal workers.

More than 50 % of the population in all the selected villages constitutes non-workers. The percentage of population engaged as agricultural labourers and cultivators is about 2 % in Vadia village and maximum in Karantha village (35.23 %). The percentage of total workers engaged in Household Industry is negligible (Table 4).

Table 4
Worker Population of the Selected Village and Rajpipla Town

(As per 2001 Census)

Particulars	*Raj pipla (M)*	*Ram para (Man grol)*	*Thari*	*Vadia (CT)-*	*Karan tha*	*Lachh ras*	*Man grol*
Total Population	34923	806	783	4480	1374	4006	1535
Total Workers	10753	375	293	1179	584	1982	609
% of Workers Population	30.79	46.53	37.42	26.32	42.50	49.48	39.67
Total Main Workers Population	10148	257	286	1004	583	1348	374
% of Main Workers to Total Population	29.06	31.89	36.53	22.41	42.43	33.65	24.36
Total Marginal Workers Population	605	118	7	175	1	634	235
% of Marginal Workers to Total Population	1.73	14.64	0.89	3.91	0.07	15.83	15.31
Total Non Workers Population	24170	431	490	3301	790	2024	926
% of Non Workers to Total Population	69.21	53.47	62.58	73.68	57.50	50.52	60.33
Total Main Agricultural Cultivator Population	914	160	240	110	484	1132	268
% of Main Agricultural Labourers, Cultivators to Total Population	2.62	19.85	30.65	2.46	35.23	28.26	17.46
Total Main Households Industry Workers Population	275	1	6	11	6	11	3
% of Main Household Industry to Total Population	0.79	0.12	0.77	0.25	0.44	0.27	0.20

MANUFACTURING IN SURENDRANAGAR DISTRICT

The total population of the district as per 2001 census is 15.15 lakhs. The proportion of rural and urban population of the total population of the district is 73.4 percent and 26.60 percent respectively. There are 9 towns in the district. The total land in the district is 10.49 lakh hectares during 1989-90. The classification of land in rural areas of the district is mentioned hereunder:

Forest land	:	30800 Hectares
Irrigated land	:	61900 Hectares
Unirrigated land	:	602500 Hectares

The area under major crops for the year 1989-90 is mentioned hereunder:

Area under all food-grain crops	:	320200 Hectares
Area under total groundnut crops	:	29600 Hectares
Area under total cotton crops	:	294400 Hectares

From the afore mentioned data, it is evident that a large area is under cotton cultivation.

The number of registered factories and average number of daily workers employed by some of the industries in the year 1989, is mentioned hereunder:

- 133 non-metallic mineral products industry employs 3203 numbers of persons per day.
- 71 establishments of Cotton textiles employ 6413 numbers of persons per day.
- 67 establishments of Machinery and machine tools and parts except electrical machinery employ 2142 number of persons per day.

Profile of Thangadh–Manufacturing Cluster

Thangadh is famous as center of Ceramic industry in India and basically a manufacturing kind of set up. This town has 1689 enterprises and the maximum number of labourers (58% of total labourers) engaged in manufacturing, processing, servicing and repairs related activities. Also looking at the commodities imported, manufactured and exported, the town seems to be manufacture dominated. There is maximum number of non-metallic mineral products industry when compared to other industries in the district. Looking at the number of enterprises, establishment of Ceramic industry and number of labourers engaged in manufacturing, processing, servicing and repair related works in Thangadh, we have selected Thangadh town as one of the three town of the State for conducting the study.

Imported: Coal, Furnace oil, Pottery material and food grains.

Manufactured: Pottery and sanitary ware, electrical goods and leather goods.

Exported: Pottery, clay and sanitary ware, crockery, electrical goods and leather goods.

Profile of Selected Villages

For conducting the study, the following six villages[2] in the vicinity of Thangadh Town has been selected:

ii.	Tarnetar	ii.	Vijalia
iii.	Manadasar	iv.	Rupavati
vi.	Songadh	vi.	Morthala

DEMOGRAPHIC PROFILE

As per the 2001 Census, the number of households in the six selected villages varies from 200 to 530 with Rupavati village having the maximum number of households i.e. 528 and Songadh village having the minimum number of households i.e. 197. The population in the selected villages varies from 1100 to 3000 with Navagam village having a maximum population of

3010 and Songadh village having the minimum population i.e. 1176 (Table 5).

Table 5
Demographic Profile of the Selected Villages and Thangadh Town

(As per 2001 Census)

Particulars	*Than gadh*	*Tar-netar*	*Vija-lia*	*Mana-dasar*	*Rupa vati*	*Son-gadh*	*Mor-thala*
Total Area of the village (Ha.)*	50.39 Sq. km.	1372.33	1340.3	978.32	2051.59	1000.83	1795.74
No. of households	6828	313	377	233	528	197	444
Total population	36880	1921	2656	1945	3010	1176	2661
Avg. size of HH	5.40	6.14	7.05	8.35	5.70	5.97	5.99
Sex Ratio	901	900	960	971	860	1007	905
Total population of children in Age group (0-6Yrs.)	6118	408	566	455	645	249	573
Population of age group (0-6 Yrs) (% of total population	17	21	21	23	21	21	22
Scheduled Caste	7253	94	147	73	41	23	201
Scheduled Caste (% total population)	20	5	6	4	1	2	8
Scheduled Tribe	53	0	14	1	8	0	6
Literates	21353	437	732	522	1040	358	839
Literates (% of Total Population)	57.90	22.75	27.56	26.84	34.55	30.44	31.53

* Census of Gujarat, 1991.

The sex ratio in the selected villages varies from 860 to 1007. Among the six villages selected, the sex ratio is lowest in Rupavati village and highest (1007) in Songadh village. The population in the age group of 0–6 years varies from 21% to 23% of the total population. The percentage of total population of Schedules Caste is maximum in Morthala village and is minimum in Rupavati village. The population of Schedules Tribes is insignificant in all the selected villages.

The literacy rates in the selected villages vary from 22.75% (Tarnetar village) to 34.55% (Navagam village).

AMENITIES

There is Primary or Elementary education facility in all the selected villages but no Matriculation and Secondary School facility available. Adult Literacy Class/ Centre exist in Morthala village.

All the selected villages have a Community Health Worker/ Health Worker or Primary Health Centre or Sub Centre. There is Child Welfare Centre in four villages i.e. Tarnetar, Vijalia, Rupavati and Morthala village. There is a family planning center in Rupavati and Morthala village.

There is potable drinking water facility in all the villages through tap water, tank water, and handpump.

Out of the six villages selected, only four villages have either post office and phone connection is available in only one village i.e. Rupavati village.

Vijalia and Manadasar village is connected by Kuchcha road and all other villages have Pucca Roads. All villages except Songadh village are connected with Bus Service.

Only two villages i.e. Manadasar and Rupavati have electricity connection for all purpose. Tarnetar, Vijalia and Songadh village have electricity connection for domestic and agriculture purpose (Table 6).

WORKER POPULATION

The percentage of population engaged as workers in the selected villages varies from 40 % to 50 % of the total population. Relative high work participation rates indicate less prosperity of the region in general.The per cent of main workers to total population varies from 27 % to 45 % with Rupavati village having the maximum percentage of population (44.39 %) engaged as main workers and Songadh village has the minimum percentage of population (27.72 %) engaged as main workers (Table 7).

Table 6
Amenities Available in the Selected Villages

(As per 2001Census)

Particulars	*Tarnetar*	*Vijalia*	*Manadasar*	*Rupavati*	*Songadh*	*Morthala*
Educational	P	P	P	P		PP, Ac (2)
Medical	CWC, CHC	CWC, SMP, CHW	CHW	CWC, PHS, FPC, CHW	CHW	CWC, PHS, FPC, CHW (2)
Drinking water (Potable)	W,TK, HP	T,W, TK, HP	T,W, TK, TW	W,TK, HP	W,TK, TW	W, TK, HP, R, N
Post & Telegraph	PO	PO	–	PO, Phone	–	PO
Day or Days Market/ Haat, if any.	–	–	–	–	–	–
Communication (Bus Stop, Railway Station, Water Way)	BS	BS	BS	BS	-	BS
Approach to village	PR, KR	KR	KR	PR, KR	PR, KR	PR, KR
Nearest town (Distance)	Than gadh (10)	Than gadh (12)	Than gadh (11)	Than gadh (9)	Than gadh (7)	Than gadh (12)
Power supply	ED, EAG	ED, EAG	EA	EA	ED, EAG	ED, EO

Among all the selected villages, Songadh village has the maximum percentage (10.88%) of population engaged as marginal workers. More than 50% of the population in all the selected villages constitutes non-workers.

Among all the six village selected, Manadasar village has maximum percentage of total population (36.45%) and Tarnetar Village has minimum percentage of total population (7.29%) engaged as Main Cultivators or Agricultural Labourers. Households dependent on agriculture are least in this cluster barring one or two villages where irrigation facility is available. Tarnetar is famous for well known mela held around Shivratri every year.

Table 7
Worker Population of the Selected Village and Thangadh Town

(As per 2001 Census)

Particulars	*Than gadh (M)*	*Vija- liya*	*Tar- netar*	*Mana- dasar*	*Mor- thala*	*Ru pavati*	*Son gadh*
Total Population	36880	2656	1921	1945	2661	3010	1176
Total Workers	12416	1280	892	901	1158	1413	454
% of Workers Population	33.67	48.19	46.43	46.32	43.52	46.94	38.61
Total Main Workers	11829	1014	719	778	1102	1336	326
% of Main Workers to Total Pop.	32.07	38.18	37.43	40.00	41.41	44.39	27.72
Total Marginal Workers	587	266	173	123	56	77	128
% of Marginal Workers to T. Pop.	1.59	10.02	9.01	6.32	2.10	2.56	10.88
Total Non Workers	24464	1376	1029	1044	1503	1597	722
% of Non Workers to Total Pop.	66.33	51.81	53.57	53.68	56.48	53.06	61.39
Total Main Agri. and Cultivator	297	843	140	709	451	551	189
% of Main Agri. Lab.&Cultivators to Total Pop.	0.81	31.74	7.29	36.45	16.95	18.31	16.07
Total Main HHs Industry Workers	253	33	0	13	21	26	0
% of Main HHs Industry workers to Total Pop.	0.69	1.24	0.00	0.67	0.79	0.86	0.00

Among all the selected villages, Songadh village has the maximum percentage (10.88 %) of population engaged as marginal workers. More than 50% of the population in all the selected villages constitutes non-workers.

Among all the six village selected, Manadasar village has maximum percentage of total population (36.45%) and Tarnetar Village has minimum percentage of total population (7.29%) engaged as Main Cultivators or Agricultural Labourers. Households dependent on agriculture are least in this cluster barring one or two villages where irrigation facility is available. Tarnetar is famous for well known mela held around Shivratri every year.

The percentage of total population engaged in Household Industry is maximum in Vijaliya village (1.24% of the total population).

COMMODITY MARKETING IN MEHSANA DISTRICT

The total population of the district as per 2001 census is 18.38 lakhs. The proportion of rural and urban population of the total population of the district is 77.60 percent and 22.40 percent respectively. The sex ratio is 927 when compared to 920 of Gujarat state. The literacy rate is 75.20 percent when compared to 69.10% of Gujarat state. Work participation rate is 45.10 % when compared to 41.90 % of Gujarat state. The total land in the district is 9.03 lakh hectares during 1989-90. The classification of land in rural areas of the district is mentioned here under: Forest Land: 14,100 Ha; Irrigated land: 3, 36,100 Ha; Unirrigated Land: 3, 59,900 Ha.

Profile of Unjha–Commodity Marketing Cluster

Unjha town is known for its Ayurvedic Pharmacy engaged in the manufacture of Ayurvedic medicines. The market yard of Unjha is the biggest in the district where a large transaction of jeera (cumin), variyali (fennel seed) and isabgul (fleawort seed) is carried out[3]. Ganj Bazaar at Unjha is the Asia's biggest spice market.

Unjha is basically a trading hub kind of set up. This town has 3763 enterprises and the maximum number of labourers (35% of total labourers) engaged in trading and commerce related activities.

Also looking at the commodities imported, manufactured and exported, the town seems to be trade dominated. In this backdrop we have selected Unjha city as one of the town for conducting the study in Gujarat State.

Imported: Calcium carbonate, foodgrains and kerosene.

Youth Spice up Trade at Unjha Masala Bazaar[4]

Dilapidated old buildings, damp and smelly dukaans, and sethias in crisp khadis. Welcome to Ganj Bazaar, Asia's biggest spice market, at Unjha. Inside the marketplace, the winds of change are blowing. So you walk into plush, air-conditioned chambers where you can spot computers and the latest cellphones.

The sethias, who have always believed in traditional ways of doing business, have brought in management trainees, geeks and Exim graduates to change things around. And the youngsters have certainly brought in change in the market which trades in at least Rs 2 crore by the end of the day's auctions.

The English-speaking graduates have introduced e-mail marketing, created websites and computerised accounting systems, and make presentations on laptops and deal directly with foreign clients. Some of them are hired by the firms while others are heirs to family businesses.

"The youngsters are doing really well in exports," says Rameshbhai Patel, an elderly trader. "Exports were earlier handled by agents in Mumbai who would deal with foreign firms. But these youngsters are turning the tables on them. They are English-speaking and know the right manners. They make presentations to clients abroad and have started dealing with them directly.

Of course, they rely heavily on our experience to strike a good deal!" he says. Patel has hired Minesh, an MCA graduate, who gave up an Aptech franchise in Vadodara for Unjha's spice market. "I have established direct contact with our foreign buyers. We send quotations on e-mail, make presentations on the Net and keep them informed about the prices on a daily basis," says Minesh.

The Ganj Bazaar has nearly 800 trading firms which deal in 27 commodities, mainly spices like jeera (cumin seeds), saunf (fennel seeds) and isabgol (psyllium). The market made a

Contd....

turnover of over Rs 900 crore in 2003. Ninety per cent of the transactions are done in cash and, as a rule, payments for stock purchase to farmers are made the same day. Spices are supplied to over 1,500 centres in India and abroad from Unjha. To handle the financial transactions, there are 17 banks in and around the market.

The switch to technical expertise has added to the profits. Patel's firm began to reap the benefits within a year of bringing in Minesh. "There is also a website and profile of the firm and its owners," says Minesh. On a foreign business tour these days, the 'seth' is usually accompanied by a hired hand or a son who would be carrying, besides samples, a laptop. And they are now travelling to as far as Sydney, South Korea, Turkey and the US to strike deals.

Manufactured: Fennel, cumin and mustard seeds, ayurvedic medicines and tooth paste.

Exported: Fennel, cumin and mustard seeds, ayurvedic medicines and tooth paste.

Profile of Selected Villages around Commodity Marketing Cluster

For conducting the study, the following six villages[5] in the vicinity of Unjha Town has been selected:

i. Jagannathpura ii. Kahoda
iii. Bhunav iv. Sunak
v. Shihi vi. Surpura

DEMOGRAPHIC PROFILE

As per the 2001 Census, the number of households in the six selected villages varies from 350 to 1400 with Jaganathpura village having the minimum number of households i.e. 357 and Kahoda village having the maximum number of households i.e. 1442. The population in the selected villages varies from 1800 to 7000 with Jagannathpura village having a minimum population of

1853 and Kahoda village having the maximum population i.e. 6978 (Table 8).

Table 8
Demographic Profile of the Selected Villages and Unjha Town

(As per 2001 Census)

Particulars	*Unjha (M)*	*Jagan-nath pura*	*Ka-hoda*	*Bhu-nav*	*Sunak*	*Shihi*	*Sur-pura*
Total Area of the village (Ha.)*	35.15 Sq. Km	464.32	1221.39	630.24	765.43	282.1	230.59
No. of households	10734	357	1442	806	699	669	393
Total population	53876	1853	6978	3809	3274	3275	2367
Avg. HH size	5.02	5.19	4.84	4.73	4.68	4.90	6.02
Sex Ratio	897	1016	981	946	1001	922	904
Total population of children in Age group (0-6 Yrs.)	5928	213	877	492	383	465	323
Population of age group (0-6Yrs) (% of total population	11.00	11.49	12.57	12.92	11.70	14.20	13.65
Scheduled Caste	4173	216	536	362	568	243	84
Scheduled Caste (% of total population)	7.75	11.66	7.68	9.50	17.35	7.42	3.55
Scheduled Tribe	367	0	10	0	0	0	0
Literates	40739	1355	4823	2561	2522	2351	1591
Literates (% of Total Population)	75.62	73.12	69.12	67.24	77.03	71.79	67.22

* As per 1991 Census.

The sex ratio in the selected villages varies from 904 to 1016. Among the six villages selected, the sex ratio is lowest in Surpura village and highest in Jaganathpura village.

The population in the age group of 0–6 years varies from 11% (Jaganathpura village) to 14% (Sunak village) of the total population.

The population of Schedules Caste in the selected villages varies from 3.5% (Surpura village) to 17% (Sunak village) of total population. The population of Schedules Tribes in the selected villages is negligible.

The literacy rates in the selected villages vary from 67.22% (Surpura village) to 77.03% (Sunak village).

AMENITIES

There is Primary or Elementary education facility and Adult Education Class/Centre in all the selected villages.

All the selected villages have a Community Health Worker/Health Worker and facility of either Primary Health Centre or Sub Centre or have Registered Private Practitioner. Maternity and Child Welfare Center is being run in only Kahoda village.

There is potable drinking water facility through tap water, well water and tube well water.

All the villages are connected through telephones and have either a post office or post and telegraph office.

All the villages have Pucca Roads and Kuchcha roads and are connected with Bus Service.

All the selected villages have electricity for all purpose (Table 9).

WORKER POPULATION

The worker population to total population varies from 40% to 64%. Relative high work participation rates indicate less prosperity of the region in general. The percentage of population engaged as main workers in the selected villages varies from 28% to 39% with Bhunav village having the maximum percentage of population (44 %) engaged as main workers and Shihi village has the minimum percentage of population (28%) engaged as main workers. The percentage of marginal workers varies from 0.50% (Jaganathpura Village) to 26% (Shihi Village). More than 35% of the population in all the selected villages constitutes of non-workers.

Table 9
Amenities Available in the Selected Villages

(As per 2001Census)

Particulars	*Jagan nath pura*	*Ka hoda*	*Bhunav*	*Shihi*	*Sunak*	*Sur pura*
Educational	P, Ac (2)	P,H, PUC, Ac(2)	P,H, Ac(2)	P,H, Ac	P,H, Ac(2)	P, Ac (2)
Medical	PHS, RP, CHW	PHC, MCW, PHC, D, FPC, RP (3), CHW, O	PHC, FPC, RP(3) CHW, O	PHS, FPC RP, CHW	RP(2) CHW	PHS,
Drinking water (Potable)	T, W, TW	T, W, TW	T, W, TW	T, W, TW	T, W, TW	T, W, TW
Post & Telegraph	PO, Phone	PTO, Phone	PTO, Phone	PO, Phone	PO, Phone	PTO, Phone
Day or Days of Market/ Haat, if any.	–	–	–	–	–	–
Communication (Bus Stop, Railway Station, Water Way)	BS	BS	BS	BS	BS	BS
Approach to village	PR, KR	PR, KR	PR, KR	PR, KR	PR, KR	PR, KR
Nearest town (Distance)	Sidh-pur (4)	Sidh-pur (8)	Sidh-pur (10)	Unjha (8)	Unjha (10)	Unjha (4)
Power supply	EA	EA	EA	EA	EA	EA

The percentage of population engaged as main agricultural labourers and cultivators varies from 26% (Shihi village) to 55% (Bhunav village). The per cent of total population engaged in Household industry is negligible in all the selected villages (Table 10).

Table 10
Worker Population of the Selected Village and Unjha Town

(As per 2001 Census)

Particulars	*Unjha (M)*	*Jagan-nath pura*	*Ka-hoda*	*Bhu-nav*	*Shihi*	*Sunak*	*Sur-pura*
Total Population	53876	1853	6978	3809	3275	3274	2367
Total Workers Population	19456	731	3559	2438	1804	1807	1342
% of Workers Population	36.11	39.45	51.00	64.01	55.08	55.19	56.70
Total Main Workers Population	17029	722	2579	1674	936	1179	768
% of Main Workers To Total Population	31.61	38.96	36.96	43.95	28.58	36.01	32.45
Total Marginal Workers Population	2427	9	980	764	868	628	574
% of Marginal Workers To Total Population	4.50	0.49	14.04	20.06	26.50	19.18	24.25
Total Non Workers Population	34420	1122	3419	1371	1471	1467	1025
% of Non Workers To Total Population	63.89	60.55	49.00	35.99	44.92	44.81	43.30
Total Main Agricultural and Cultivator Population	1331	405	1342	1357	478	705	390
% of Main Agricultural Labourers and Cultivators To Total Population	6.84	55.40	37.71	55.66	26.50	39.01	29.06
Total Main Households Industry Workers Population	153	12	45	19	7	29	0
% of Main Household IndustryTo Total Population	0.79	1.64	1.26	0.78	0.39	1.60	0.00

Notes

1. All the Selected Villages are within the Range of 5 to 15 km surrounding Rajpipla Town.
2. All the Selected Villages are within the Range of 5 to 15 km surrounding Thangadh Town.
3. http://www.all-indiatravel.com/unjha-zone/
4. The Indian Express, June 07, 2004
5. All the Selected Villages are within the Range of 5 to 20 km Surrounding Unjha Town.

5

FIELD SURVEY AND RESULTS

In accordance with the methodology chosen, field survey was conducted covering the urban areas and the six villages in each of their vicinity. Profile of sample households in each cluster of the villages are presented in Tables 1, 2 and 3.

The high % of agriculture and agricultural labour Households indicate the predominance of dependence on agriculture in this cluster. Low level of jobs and trade and services in the villages also indicate the same.

The low % of agriculture and agricultural labour Households indicate the relative low importance of agriculture in this cluster. Non-agricultural labour, trade and services and jobs are relatively more important, even though all of them are of very low skilled.

The high % of agriculture and labour Households indicate the predominance of dependence on agriculture and commodity trade in this cluster. Low level of jobs and labour in the villages also indicate the same.

Demographic features of the sample Households in terms of composition of family members, average size of the HH, sex ratio, and dependent child population is presented below:

Sex ratio in both rural and urban areas is very imbalanced in the age groups below 55. Sex ratio in the

Table 1
Profile of Sample Households - Rajpipla –Trade & Services Cluster villages

Categories	*Mongrol*		*Vavdi*		*Thari*		*Karanta*		*Lacchras*		*Rampura*		*Total*	
	No	*%*	*No*	*%*	*No*	*%*	*No*	*%*	*No*	*%*	*No*	*%*	*No*	*%*
Agriculture	11	36.7	6	19.4	16	53.3	11	37.9	16	53.3	0	0.0	60	33.3
Tr.&Service	s	6.7	3	9.7	1	3.3	2	6.9	2	6.7	4	13.3	14	7.8
Ag. Labour	10	33.3	0	0.0	1	3.3	0	0.0	0	0.0	22	73.3	33	18.3
Labour	4	13.3	13	41.9	9	30.0	16	55.2	11	36.7	4	13.3	57	31.7
Jobs	3	10.0	9	29.0	3	10.0	0	0.0	1	3.3	0	0.0	16	8.9
Total	30	100.	31	100.	30	100.	29	100.	30	100.	30	100.	180	100.0

Table 2
Profile of Sample Households – Thangad Industrial cluster villages

Categories	Rupavati		Morthala		Manadsar		Songadh		Vijiyala		Tarnetar		Total	
	No	*%*	*No*	*%*	*No*	*%*	*No*	*%*	*No*	*%*	*No*	*%*	*No*	*%*
Agriculture	7	26.9	17	58.6	16	55.2	13	43.3	19	63.3	18	62.1	90	52.3
Tr.&Services	5	19.2	2	6.9	2	6.9	0	0.0	2	6.7	0	0.0	11	6.4
Ag. Labour	3	11.5	1	3.4	2	6.9	1	3.3	0	0.0	2	6.9	9	5.2
Labour	6	23.1	9	31.0	9	31.0	15	50.0	8	26.7	7	24.1	54	31.4
Jobs	5	19.2	0	0.0	0	0.0	1	3.3	0	0.0	2	6.9	8	4.7
Total	26	100.0	29	100.	29	100.0	30	100.	30	100.	29	100.	172	100.0

Table 3
Profile of Sample Households–Unjha Commodity Market cluster villages

Categories	Surpura		Kahoda		Bhunav		Sunak		JN Pura		Shihi		Total	
	No	*%*	*No*	*%*	*No*	*%*	*No*	*%*	*No*	*%*	*No*	*%*	*No*	*%*
Agriculture	16	53.3	13	43.3	8	25.8	14	46.7	23	76.7	18	60.0	92	50.8
Tr.&Services	2	6.7	6	20.0	6	19.4	5	16.7	6	20.0	1	3.3	26	14.4
Ag. Labour	4	13.3	1	3.3	11	35.5	2	6.7	1	3.3	1	3.3	20	11.0
Labour	5	16.7	6	20.0	3	9.7	7	23.3	0	0.0	10	33.3	31	17.1
Jobs	3	10.0	4	13.3	3	9.7	2	6.7	0	0.0	0	0.0	12	6.6
Total	30	100.0	30	100.0	31	100.0	30	100.0	30	100.0	30	100.0	181	100.0

Table 4

Demographic Profile of Sample Households in Towns & Villages – Rajpipla Cluster

	Composition in Sample HHs											
	<15		*15-55*		*>55*		*Sample HHs*	*Total in HHs*		*Total Persons*	*AVG HHs*	
Particulars	*M*	*F*	*M*	*F*	*M*	*F*	*HHs*	*M*	*F*	*in HHs*	*Size*	*<15 as %*
Narmada-Rajpipla	72	68	180	144	26	38	120	278	250	528	4.40	26.51
Urban sex ratio	944		800		1461			899				
Mongrol	21	17	53	34	11	13	30	85	64	149	4.97	25.50
Vavdi	34	19	53	34	15	19	31	102	72	174	5.61	30.46
Thari	24	18	40	25	10	14	30	74	57	131	4.37	32.06
Karanta	24	12	53	30	12	17	29	89	59	148	5.10	24.32
Lachhras	18	17	47	32	14	16	30	79	65	144	4.80	24.31
Rampura	28	29	48	34	7	8	30	83	71	154	5.13	37.01
Rural Area Total	149	112	294	189	69	87	180	552	388	900	5.00	29.00
Rural sex ratio	752		629		1261			703				

Table 5

Demographic Profile of Sample Households in Towns & Villages- Thangad Cluster

Particulars	*Composition in Sample HHs* *<15* *M*	*F*	*15-55* *M*	*F*	*>55* *M*	*F*	*Sample HHs* *HHs*	*Total in HHs* *M*	*F*	*Total Persons in HHs*	*AVG HHs Size*	*<15 as %*
Surendranagar-Thangad	130	90	180	135	23	27	124	341	252	593	4.78	37.10
Urban sex ratio	692		750		1174			739				
Rupavati	46	49	38	34	2	5	29	86	88	174	6.00	54.60
Morthala	45	28	44	26	3	3	30	92	57	149	4.97	48.99
Manadasar	36	33	51	29	5	9	30	92	71	163	5.43	42.33
Songadh	26	25	50	31	2	0	30	78	56	134	4.47	38.06
Vigiyala	34	19	52	34	1	6	30	87	59	146	4.87	36.30
Tarnetar	45	29	50	32	5	7	30	100	68	168	5.60	44.05
Rural Area Total	232	183	285	186	18	30	179	535	399	934	5.22	44.43
Rural sex ratio	789		653		1666			746				

Table 6

Demographic Profile of Sample Households in Towns & Villages – Unjha Cluster

Particulars	*Composition in Sample HHs* <15 M	F	15-55 M	F	>55 M	F	*Sample HHs* HHs	*Total in HHs* M	F	*Total Persons* in HHs	*AVG HHs* Size	<15 as %
Mehsana-Unjha	121	79	204	154	37	44	127	362	277	639	5.03	28.17
Urban sex ratio	653		755		1189			765				
Surpura	28	24	49	36	17	14	30	94	74	168	5.60	30.95
Kahoda	23	21	57	31	16	14	31	96	66	162	5.23	27.16
Bhunav	31	24	59	43	12	10	31	102	77	179	5.77	30.73
Sunak	20	15	43	39	7	9	30	70	63	133	4.43	26.32
JNPura	33	20	46	48	13	13	30	92	81	173	5.77	30.64
Shihi	23	21	53	36	15	16	30	91	73	164	5.47	26.83
Rural Area Total	158	125	307	233	80	76	182	545	434	979	5.38	28.90
Rural sex ratio	791		759		950			796				

age group of persons above 55 is in favour of women. The sex ratio across all the age groups both in rural and urban areas in Rajpipla cluster is very imbalanced.

Sex ratio in both rural and urban areas is very imbalanced in the age groups below 55 in this cluster of villages also. The average family size both in urban and rural areas is higher and also the numbers of children below 15 who are dependent are more in the sample Households in Thangad cluster. Sex ratio in the age group of persons above 55 is in favour of women.

Sex ratio in both rural and urban areas is very imbalanced in the age groups below 55 in both urban and rural in Unjha cluster also. The average family size both in urban and rural areas is closer and also the numbers of children below 15 who are dependent are more in the sample Households in Thangad cluster. Sex ratio in the age group of persons above 55 is in favour of women.

Housing and amenities within the house and access to drinking water and sanitation facilities are presentedin Tables 7, 8 and 9.

Table 7
Housing, Amenities in Sample Households in Towns & Villages Rajpipla Cluster

Particulars	*Kaccha*	*Pucca*	*Own*	*Hired*	*Elec-trified*	*Drinking Water*	*With Toilet*
Narmada-Rajpipla	32	88	104	16	119	120	113
Urban Area (%)	26.67	73.33	86.67	13.33	99.17	100	94.17
Mongrol	19	11	28	2	30	30	18
Vavdi	17	14	28	3	31	31	17
Thari	17	13	29	1	30	30	19
Karanta	14	15	27	2	29	29	13
Lachhras	15	15	29	1	30	30	18
Rampura	28	2	30	0	30	30	1
Rural Area Total	110	70	171	9	180	180	86
Rural Area (%)	61.11	38.39	95.0	5.0	100	100	47.78

In Rajpipla cluster of villages, significant number (61.11 %) of Households are kaccha houses and 52.22% of Households are still without sanitation facilities.

Table 8
Housing and Amenities in Sample Households in Towns & Villages Thangad Cluster

Particulars	*Kaccha*	*Pucca*	*Own*	*Hired*	*Elec-trified*	*Drinking Water*	*With Toilet*
Surendranagar-Than	39	82	97	24	121	121	79
Urban Area (%)	32.23	67.77	80.17	19.83	100	100	65.29
Rupavati	22	8	29	1	17	30	2
Morthala	23	6	27	2	3	30	1
Manadasar	23	6	27	2	15	29	0
Songadh	25	6	24	7	6	30	5
Vigiyala	27	3	28	2	18	30	3
Tarnetar	29	1	29	1	1	30	0
Rural Area Total	149	30	164	15	60	179	11
Rural Area (%)	83.24	16.76	91.62	8.38	33.52	100	6.15

Highest number of kaccha houses (83.24%) and low electrification (33.52%) in Thangad cluster of villages may be noted. However, in reality many of the other houses have illegal and unauthorized connections and do not pay bills. In this cluster of villages, significant numbers of Households (93.85 %) are without sanitation facilities.Even in urban area 34.71% Households are still without access to sanitation facilities. In some of the villages houses with electricity and sanitation facilities are very few. In terms of urban rural disparity in housing and amenities this cluster of villages are very poor.

In comparison significant numbers of rural Households in Unjha cluster have better housing (58.80% pucca), amenities within the house and access to sanitation facilities (60.44%). Still many houses (39.56%) do not have access to sanitation facilities.

Income and sources of income are presented in Tables 5.10, 5.11 and 5.12. Rajpipla cluster of villages has the highest average income per HH compared with villages in the other clusters. Wages 1 represent wages earned within the village and Wages 2 represent wages earned outside the village.

Table 9
Housing and Amenities in Sample Households in Towns & Villages Unjha Cluster

Particulars	*Kaccha*	*Pucca*	*Own*	*Hired*	*Elec-trified*	*Drinking Water*	*With Toilet*
Mehsana-Unjha	16	108	108	16	124	124	117
Urban Area (%)	12.90	87.10	87.10	12.90	100	100	94.35
Surpura	17	13	30	0	21	30	12
Kahoda	4	27	31	0	30	31	27
Bhunav	14	17	25	6	28	31	17
Sunak	20	10	27	3	29	30	15
JNPura	8	22	30	0	27	30	24
Shihi	12	18	30	0	30	30	15
Rural Area Total	75	107	173	9	165	182	110
Rural Area (%)	41.20	58.80	95.05	4.95	90.66	100	60.44

In Rampura village Households are significantly dependent on dairying and wages for income. All the Households in the other villages are largely dependent on agriculture for income. In Mongrol which is nearer to Rajpipla some people get wage employment in urban area.

In Thagngad cluster Households are more dependent on wages within and outside the villages and relatively less on agriculture barring Songadh. Dairying also brings significant share of income for the Households. Due to unskilled nature of clay mining jobs and jobs in the ceramics industry, rain dependent agriculture and low rain fall average income of the Households is less and the corresponding quality of life. Working conditions in the

Table 10

Income Particulars of Sample Households Rajpipla Cluster

Village	*Agriculture*	*Wages1*	*Wages 2*	*Dairy*	*Others*	*Total Income*	*AVG HHIncome*
Mongrol %	58.27	10.44	12.99	17.48	0.82	100.00	Mongrol Total
1874000	335600	417600	562200	26500	3215900	107197	
Vavdi %	42.61	14.78	0.43	11.16	31.03	100.00	
Vavdi Total	1196000	414700	12000	313200	870800	2806700	90539
Thari %	78.54	5.34	0.82	12.20	3.09	100.00	
Thari Total	2746000	186800	28800	426600	108000	3496200	116540
Karanta %	65.22	12.13	0.00	22.66	0.00	100.00	
KarantaTotal	1423000	264600	0	494400	0	2182000	75241
Lacchras %	76.59	10.69	0.00	12.71	0.00	100.00	
Lacchras Total	1637500	228600	0	271800	0	2137900	71263
Rampura %	12.38	31.08	1.64	54.89	0.00	100.00	
Rampura Total	49000	123000	6500	217200	0	395700	13190
Rural Area Total	8925500	1553300	464900	2285400	1005300	14234400	79080
Rural Area (%)	62.70	10.91	3.27	16.06	7.06	100	

Table 11
Income Particulars of Sample Households Thangad Cluster

Village	*Agriculture*	*Wages1*	*Wages 2*	*Dairy*	*Others*		*HH Income*
Rupavati %	11.58	36.50	17.59	14.30	20.03	100.00	
Rupavati Total	102000	321600	155000	126000	176500	881100	33888
Morthala %	23.71	36.06	10.41	24.04	5.79	100.00	
Morthala Total	213000	324000	93500	216000	52000	898500	30983
Manadsar %	17.74	28.44	14.22	30.28	9.33	100.00	
Manadsar Total	232000	372000	186000	396000	122000	1308000	45103
Songadh %	40.57	13.41	20.49	7.66	17.87	100.00	
Songadh Total	635500	210000	321000	120000	280000	1566500	52217
Vijiyala %	28.56	21.40	4.88	37.04	8.12	100.00	
Vijiyala Total	374700	280800	64000	486000	106500	1312000	43733
Tarnetar %	13.08	44.98	16.19	25.36	0.40	100.00	
Tarnetar Total	164000	564000	203000	318000	5000	1254000	43241
Rural Area Total	1721200	2072400	1022500	1662000	742000	7220100	41977
Rural Area (%)	23.84	28.70	14.16	23.02	10.28	100	

Table 12

Income Particulars of Sample Households Unjha Cluster

Village	*Agriculture*	*Wages1*	*Wages 2*	*Dairy*	*Others*		*HH Income*
Surpura %	12.27	28.49	0.00	55.35	3.89	100.00	
Surpura Total	315200	732000	0	1422180	100000	2569380	85646
Kahoda %	24.55	12.26	1.10	61.59	0.50	100.00	
Kahoda Total	490000	244700	22000	1229184	10000	1995884	66529
Bhunav %	5.39	43.29	2.64	47.69	0.99	100.00	
Bhunav Total	147000	1180200	72000	1300000	27000	2726200	87942
Sunak %	16.43	12.90	6.80	62.01	1.87	100.00	
Sunak Total	366200	287400	151500	1381992	41600	2228692	74290
JNPura %	37.35	7.51	0.00	55.05	0.09	100.00	
JNPura Total	830000	166800	0	1223196	2000	2221996	74067
Shihi %	7.74	25.58	1.48	64.97	0.24	100.00	
Shihi Total	169200	559272	32300	1420800	5200	2186772	72892
Rural Area Total	2317600	3170372	277800	7977352	185800	13928924	76955
Rural Area %	50.92	23.49	6.10	15.41	4.08	100	

ceramic industry are poor and the waste from the industry is increasingly filling the waste and not so waste lands in the vicinity.

Composition of the sample Households in three towns selected for the study are presented in Table 13.

Table 13
Composition of Sample Households in the Towns

Category	*Rajpipla*		*Thangad*		*Unjha*		*Total*	
	No	*%*	*No*	*%*	*No*	*%*	*No*	*%*
Agriculturists	9	7.50	0	0.00	13	10.24	22	5.93
Business	16	13.33	13	10.48	28	22.05	57	15.36
Shops	19	15.83	28	22.58	11	8.66	58	15.63
Services	12	10.00	35	28.23	11	8.66	58	15.63
Jobs	41	34.17	6	4.84	17	13.39	64	17.25
Private Jobs	16	13.33	1	0.81	26	20.47	43	11.59
Labour	6	5.00	39	31.45	18	14.17	63	16.98
Others	1	0.83	2	1.61	3	2.36	6	1.62
Total	120	100.00	124	100.00	127	100.00	371	100.00

Table 14
Aggregate Income of Sample households in Towns

Category	*Rajpipla Income*	*Thangad Income*	*Unjha Income*	*Urban Areas Total Income*
Agriculture	461200	0	208496	669696
Households	9	0	8	17
AVG Income	51244	0	26062	39394
Business	1241500	616000	1087000	2944500
Households	16	13	16	45
AVG Income	77594	47385	67937	65433
Shops	992700	1619100	1210000	3821800
Households	19	14	9	42
AVG Income	52247	115650	134444	90995
Services	794400	1457550	458400	2710350
Households	12	30	7	49
AVG Income	66200	48585	65485	55313
Jobs	3948900	530000	1223500	5702400
Households	41	6	14	61

Contd....

Category	*Rajpipla Income*	*Thangad Income*	*Unjha Income*	*Urban Areas Total Income*
AVG Income	96315	88333	87393	93482
Private Jobs	894100	84000	965600	1943700
Households	16	1	19	36
AVG Income	55881	84000	50821	53992
Labour	62500	619750	440500	1122750
Households	6	33	13	52
AVG Income	10417	18750	33885	21591
Others	72000	43700	50000	165700
Households	1	1	1	3
AVG Income	72000	43700	50000	55233
Urban Area Total	8467300	4970100	5643496	19080896
Households	120	98	87	305
AVG Income	70561	50715	64868	62560

Average income among the sample households in the rural areas is highest in Rajpipla trade and services cluster closely followed by Unjha commodity marketing cluster.

Agriculture and dairying seem to contribute to the rural wealth significantly. Poor agriculture is the cause of low rural incomes in Thangad cluster of villages. Even in this cluster dairying seems to have an ameliorating effect. Variation of average income among the sample households in the rural areas in Rajpipla trade and service cluster is high compared to rural areas in other clusters.

Average income among the sample households in the urban areas is highest in Rajpipla town closely followed by Unjha. Households dependent on agriculture, business, and jobs seem to be doing well in Rajpila town by Unjha town. Households dependent on shops in Unjha and Thangad towns are having better incomes than those in Rajpipla.

6

GUJARAT STUDY RESULTS

AN ECONOMY IN TRANSITION

During the year 2002-03, the growth of GSDP at constant prices has been estimated at 2.1 per cent as against 9.5 per cent growth recorded in previous year. The growth in GSDP in 2002-03 decreased as compared with previous year due to lower agriculture production. The primary sector has registered negative growth with 9.8 per cent, while secondary and tertiary sectors have registered positive growth with 4.4 percent and 5.6 per cent respectively in 2002-03, as against the growth of 34.9 per cent, 0.4 per cent and 8.7 per cent registered in the previous year (2001-02) for respective sectors.

Gujarat Has Highly Diversified Sources of Income

About 78 per cent of the State Domestic Product in Gujarat comes from non-primary sectors, as against 60 per cent of the national economy. Similarly, Gujarat has a relatively diversified workforce, with more than 40 per cent employed in non-primary sectors, as against 32.6 per cent of the all-India workforce. The money and capital markets in the state are well developed, allowing for a relatively easy access to finance for its enterprising population.

IN THE FOOT STEPS OF DEVELOPD ECONOMIES

The major development was the restructuring of the industrial sector towards chemical based industries in the Vadodara–Valsad belt that is called the Golden Corridor. There was also an overall improvement in physical infrastructure, especially in Saurashtra- Kutch region where clusters had come in the late 1980s. This includes the ship-breaking yard near Bhavnagar, the diamond industry in Amreli and Bhavnagar and mineral based industries in coastal Saurashtra, in addition to the already existing developed clusters of engineering industry in Rajkot. Also there was a boom in the construction industry, especially in the urban areas. It can be concluded that Gujarat is following a development pattern of developed economies where the economy becomes highly industrialized before the service sector expands.

Increasing Rural–Urban Divide

A striking feature of sectoral incomes in Gujarat is the highly fluctuating character of per capita income from the primary sector, which varies from Rs. 792 in 1980-81 to Rs. 729 in 1990-91 to Rs. 896 in 1994-95 and to Rs. 491 in 2000-2001 (Table 3). Over a period of two decades i.e. from 1981 to 2001, the per capita income has decreased by 38 percent in case of primary sector where as in case of secondary and tertiary sector the per capita increased by 131 percent and 113 per cent respectively. On the whole, Gujarat's economy has undergone significant structural transformation with the share of the non-primary sectors having significantly increased. Also, agriculture in the state is highly unstable, which tends to leave the small farmers and farm labour in poverty. (Hirway and Mahadevia, 2004: 25).

Urban Poverty

Urban poverty is a relatively serious problem in the state, with the incidence of urban poverty being higher than that of rural poverty. The incidence of urban poverty declined only marginally during the 1980s, with the absolute number of the urban poor increasing from 41.33 lakhs in 1977-78 to 47.26 lakhs in 1983, and to 52.63 lakhs in 1987-88. The probable reasons cited for this were (a) Decline in the organized textile industry (the economy of some small towns was centred around a composite textile mill or two) leading to direct and indirect unemployment (b) Declining quality of employment arising from informalisation of the labour market (c) Urbanisation as a result of migration from rural areas. There has been a significant decline in the incidence of urban poverty in Gujarat, thanks to the increasing industrial investment. The incidence of poverty, however, is higher in smaller towns, largely because of the low employment avenues, and the predominance of low skills/productivity of work, low wages and poor infrastructural facilities in smaller towns.

The urban poverty was higher than the rural poverty in all regions of the state except in the southern plains. This decline in rural poverty could be due to sectoral diversification in rural areas of the state, which is reflected in non-household based manufacturing and other non-farm activities (Basant et al., 1998). This could also be inferred as successful implementation of anti-poverty programmes, which might have more successful in generating income in rural areas. Also Gujarat being one among the worst affected states during droughts has been receiving funds under the drought relief programmes. Also a large segment of the poor migrating from the rural areas to the urban areas are getting absorbed in low productivity tertiary activities where employment has increas in the recent years (Kundu 2000, 3117).

Self-employment and Casualisation of Labour

The share of self-employed is about 59 per cent in rural areas and 41 per cent in urban areas (1999-00). The share of casual workers has increased from 18.3 per cent to 24.9 per cent in urban areas and has declined only marginally, from 34.8 per cent in 1983 to 34.5 per cent in 1999-00 in rural areas with the share fluctuating during the period. Compared to other states, the share of casual labour is much higher in Gujarat (Hirway and Mahadevia, 2004: 34).

Only 6.3 per cent of workers are regularly employed in rural areas, while the corresponding figure in urban areas is 34.1 per cent. This share declined significantly during the 1990s in both rural and urban areas. While overall employment in the state has improved, the share of regular employment has declined, particularly in the 1990s, and the share of casual employment has increased. The share of workers covered under social security schemes of the organized sector is much less than 10 per cent. (Hirway and Mahadevia, 2004: 35).

Pollution and Degradation of Natural Resources

Pollution-prone industries are predominant in Gujarat partly because of some historical reasons, such as the growth of textile-related chemical industries, discovery of oil and natural gas, and the growth of drug and pharmaceuticals industries in the state in the pre and post-independence period. Recently, there has been a quantum jump in mineral- based industries as well. However, the pollution generated by these industries has not been controlled adequately, with the result that the workers of these units as well as the people living in surrounding areas suffer from serious health hazards. Although the state government has initiated efforts in promoting knowledge-based industries recently, not much has been achieved so far. It is also relevant to see who pays the cost of the pollution. Studies have shown that the victims of the adverse impact of pollution are industrial workers, particularly those working in small-scale units as casual and contract workers, and the vulnerable population in the surrounding areas.

RURAL URBAN DISPARITY ACROSS CLUSTERS

1. Sex ratio in all the three urban areas and the six villages selected in each of their vicinity is highly imbalanced across all age groups below 55 and across all age groups put together. In some villages and urban areas sex ratio is in favour of women above age group 55.
2. Literacy in Thangad cluster of villages is much less compared to villages in other clusters. Primary education and access to health care facilities, though available in many cases, the quality of service and actual access for use are not satisfactory especially in the villages. All villages excepting a few are connected by pucca road and a few villages still do not have phone and communication facilities.
3. Housing and amenities in the house–electricity, access to drinking water–are better in Rajpipla trade and services cluster and Unjha commodity marketing cluster compared to Thangad manufacturing and industrial cluster.
4. Average income among the sample households in the rural areas is highest in Rajpipla trade and services cluster closely followed by Unjha commodity marketing cluster. Agriculture and dairying seem to contribute to the rural wealth significantly. Poor agriculture is the cause of low rural incomes in Thangad cluster of villages. Even in this cluster

dairying seems to have an ameliorating effect. Variation of average income among the sample households in the rural areas in Rajpipla trade and service cluster is high compared to rural areas in other clusters.

5. Average income among the sample households in the urban areas is highest in Rajpipla town closely followed by Unjha. Households dependent on agriculture, business, and jobs seem to be doing well in Rajpila town by Unjha town. Households dependent on shops in Unjha and Thangad towns are having better incomes than those in Rajpipla.

MACRO ECONOMIC POLICY ISSUES

Widening Gap between Rich and Poor

Negative impacts of liberalization policies were recognized soon after launching of these policies and the need for ameliorating measures was recognized by the efforts to give it human face. The processes of globalization and economic liberalization ushered in the early nineties have contributed to the acceleration of economic growth. But they also appeared to have resulted in widening the gap between the rich and poor. They also seem to be contributing to further regional inequality, family level food insecurity, and unemployment. Liberalization has resulted in a greater degree of skewness in the pattern of development, perhaps leading to the marginalization of some states in the overall dispensation. The problems and constraints associated with agriculture sector seem to have worsened under the reforms regime.

Negligence of Social Objectives in Reforms

Quality and security of rural livelihoods seems to have been adversely affected with the low level of spending by both center and states on social sectors including health, family welfare and education. If access to food, education, technology, credit, insurance, information and natural resources like land, water and forests is going to mediated by market forces alone, the lot of the economically marginalized with very little purchasing power at their disposal will become more precarious. Both first and second generation reforms so far overlook various social objectives such as eradication of poverty, prevention of concentration of economic power in fewer hands, regional imbalances and promotion of labour intensive economic activities creating more job opportunities.

Increasing Food Stocks and Decreasing Purchasing Power

The arguments for a universal, not targeted, National Rural Employment Guarantee Act as well as for a universal Public Distribution System (PDS) are far stronger than most people realise. Rural India is in deep and continuing distress. Unemployment continues to rise. Output growth continues to fall; last year's gross food grains output were 205 million tonnes or only 162 kg net output per head of population. Loss of purchasing power owing to drastic reductions in the State's spending on rural development during the last decade and slow or no growth in agriculture continues and is reflected in a steep fall in per head food grains absorption which is now one of the lowest in the world at around 154 kg, for all-India, 20 kg lower than a mere six years ago, and it is lower still in village India where calorie intake per head continues to decline.

Forty years of successful effort to raise food grains absorption through the Green Revolution and planned expansionary policies has been wiped out in a single decade of deflationary economic reforms and India is back to the food grains availability level of 50 years ago. The farmers of Punjab and Haryana are in crisis for they have lost internal markets to the tune of 26 million tonnes of food grains—for if grain absorption had been maintained at least at the 1991 level of 178 kg per head instead of falling, the internal demand would have been 26 million tonnes higher than it actually is today.

Deepening Crisis in Rural and Agrarian Sectors

According to NSS data, five years back in 2000 there were three States where more than one-third of the rural population was below 1800 calories daily energy intake, while today the latest unpublished NSS data, according to a recent public lecture by Prof. V.S. Vyas, show eight major States to be in that position. There is already a sub-Saharan Africa (SSA) within India—half of our rural population or over 350 million people are below the average food energy intake of SSA countries.

Of the rural population, 75 per cent was actually poor in 1999-2000, being below 2400 calories required daily intake, but the Planning Commission gave us an unrealistic estimate of only 27 per cent by applying a poverty line which is far too low at less than Rs.11 a day, at which less than 1900 calories could be accessed. Rather than start food- for-work on a larger scale to generate employment, it used exports and open market sales to get rid of most of the 40 million tonnes of excess stocks, which had built up owing to more and more empty stomachs over four preceding years, as falling purchasing power prevented the poor from buying enough food, and as the actually poor were denied BPL cards because they were not officially deemed to be poor.

This experience has shown that more exports are easy — simply let your own population go hungry by following income-deflating macro policies and by 'targeting' the PDS, and apparent 'surplus' for export automatically emerges.

The exports were at below BPL price and were mostly used as animal feed abroad.

Decreasing Rural Development Expenditures

As far back as the 7th Plan, 1985-90, when Rajiv Gandhi was the Prime Minister, the average share of rural development expenditures in Net National Product (NNP) had been raised to nearly 4 per cent, with very positive effects by way of rising employment and rising real wages. In rural development expenditures five heads—agriculture, rural development, irrigation and flood control, special areas programmes, and village and small scale industry–are included.

It might be of interest to know that in the first year of UPA rule, rural development expenditures taking these same heads have been slashed to an all-time low of 0.6 per cent of NNP. The absolute outlay (budgetary estimate) was a paltry Rs.13.5 thousand crore as anyone may check from the latest Economic Survey, and this sum is actually the same, even without any adjustment for price rise, as that spent under Rajiv Gandhi 15 years ago in 1989-90. This brutal contraction in spending last year has added to the present continuing crisis of jobs and has increased hunger in rural areas.

Union Finance Minister P. Chidambaram has acted as though the agrarian crisis simply did not exist. He has signalled clearly that his priority is to appease finance and adhere to the arbitrary spending cuts spelt out in the Fiscal Responsibility and Budgetary Management Act, and not to address rising unemployment and hunger. The contrast could not be more marked, with the recent actions of the governments of France and Germany, which have relaxed fiscal austerity norms to deal with much milder unemployment problems than we face today in India.

Abolish Arbitrary Division between Above and Below Poverty

Rural India is in acute distress and the distress is bound to turn to turmoil if its crisis is not addressed. It is not too late. The need of the hour is to implement immediately and with sincerity a demand driven, universal employment guarantee and at the same time to abolish the arbitrary division between the `above' and `below' poverty line population to allow the poor to access affordable food.

Rural India Needs a New Deal

Although they may not yet be aware of it, the Prime Minister and Finance Minister today have a historic choice before them. How will history remember them: will they take the disastrous path of Bruning's deflation, or the virtuous path of Roosevelt's New Deal? Chancellor Heinrich Bruning of Germany, nicknamed the `Hunger Chancellor' had continued to follow deflationary policies even when unemployment was already high, in order to appease international financial interests. The resulting even higher unemployment discredited the Weimar Republic and made the rise to power of fascism a certainty. President Roosevelt, on the other hand, following what came to be known as Keynesian policies, offered expansionary macroeconomic measures to lift the U.S. out of depression. What India needs today is a genuine new deal for the rural poor. All those who understand the urgency of the issue of agrarian distress in this country today would wish to convey a sense of that urgency to the Government. It is time for Kumbhakarna to wake up.

References

Alagh, Y.K. (1995a), 'Gujarat, the Decisive Phase', Anvesak, 25 (1): 116-23.

Basant, R. (1993), 'Diversification of Economic Activities in Rural Areas: Key Results of a Filed Survey'. Working paper No. 53. Ahmedabad: Gujarat Institute of Development Research.

Basant, R., B.L. Kumar and R. Parthasarathy. (1998), Non-Agricultural Employment in Rural India. Jaipur: Rawat Publications.

Government of Gujarat (2000), Socio Economic Review Gujarat State (1999-2000). Gandhinagar: Directorate of Economics and Statistics, Gandhinagar.

Government of Gujarat (2000), Report on Fourth Economic Census, 1998 Gujarat. Vol. I & II. *Socio Economic Review* Gujarat State (1999-2000). Gandhinagar: Directorate of Economics and Statistics, Gandhinagar.

Hirway, I. (2000), 'Dynamics of Development in Gujarat: Some Issues', *Economic and Political Weekly*, XXXV (35 & 36): 3106-20.

Kundu, A. (2000), 'Globalization Gujarat: Urbanization, Employment and Poverty', *Economic and Political Weekly*, XXXV, (35 & 36): 3172-82.

NIRD (2003), Rural Development Statistics. Hyderabad: National Institute of Rural Development (NIRD).

Rani, Uma. (2004), 'Core Issues in the Economy and Society of Gujarat'. From In Search of Sustainable Livelihood Systems, Ed. Ruedi Baumgartner, Ruedi Hogger, New Delhi: Sage Publications India Pvt. Ltd.

INDEX